ASTRONAVIGATION

from Columbus to William Barentsz for the modern yachtsman

ASTRONAVIGATION

from Columbus to William Barentsz for the modern yachtsman

with four-year solar declination tables
day length tables for all latitudes
and the Regiment of the North Star

Siebren van der Werf und Dick Huges

Lanasta

- The cover photo was taken on Bear Island on 13 June 2011, 415 years to the day after William Barentsz left the island. The day before their departure, the encounter with a polar bear had taken place, to which the island owes its name, given by Barentsz. Photo: SvdW

- Original title: *Astronavigatie van Columbus tot Willem Barentz*

  Translation: Siebren van der Werf & Dick Huges

ISBN: 978-90-8616-342-7

www.lanasta.com

NUR: 466

1st Print, Oct. 2021

Layout:
Jantinus Mulder

Lanasta, Oude Kampenweg 29, 7873 AG Odoorn (NL).

CONTENTS

Betelgeuse

Bellatrix

Belt of Orion

Orion

Rigel

Preface

Sixteenth century handbooks for navigators, in particular the earliest of them, usually covered three subjects. In the first place there was an introduction to cosmography. The Earth was the center of the universe. It was spherical, surrounded by water, higher up by air and still higher up by a ring of fire. Around it moved on circular orbits the moon, the sun, the planets and the stars, each fixed on its own round-spinning sphere. Above it all were the heavens.
There was a section that explained how you could find your latitude by measuring the altitude of Polaris, the "Regiment of the North Star". Thirdly of course, instructions were given for taking the sun's height at meridian transit, when you see it at its highest point, directly south of you, or north if you cruise the southern waters. You will need the sun's declination and the "Regiment of the astrolabe and the quadrant" explained the sight reduction that should give you your latitude.
In this booklet we revive this format that served the navigators of the fifteenth and sixteenth centuries, from Christopher Columbus to William Barentsz. We add a simple method by which you can get a reasonably accurate guess of your longitude by timing the moment of sunset or sunrise. Our forefathers had sandglasses, which were not precise enough for this purpose, but the modern yachtsman has an accurate time keeper. Even a cheap little quartz clock will do. For the sun we give declination tables for four years, computed for 2021 - 2024, just over 500 years after the first printed set of 4 year tables in the Portuguese "Regiment of Évora", which gave them for 1517 - 1520. Our tables are organized similarly: one declination per day, here given for mid-day on the Greenwich meridian, at 12:00 GMT.
The simple questions a navigator has to answer are "where am I?" and "which way do I go from here?". Today we read the answers to both questions from a screen that translates for us the information it receives from artificial satellites. Whereas in the olden days the navigator himself had to do the sight reduction from measured altitude to position, data reduction is now done by a software program.
The instruments for measuring altitudes in the sixteenth century were the quadrant, the astrolabe and the cross-staff. The modern yachtsman has a sextant and of course an accurate clock.
The reasons for taking an interest in celestial sextant navigation are different from the time when it was a necessity. In the first place, there is the fascination and satisfaction that a sailor feels when he so finds his position. And are

we not all romantics at heart? Yes, of course, but we are also realists. Having a backup for the eventuality that your electronics will let you down and your screens go blank, is a sensible thing. You will still want to find your port.
The last chapter describes, as an illustration, a trans-Atlantic crossing by one of the authors (DH) in 1984, when he used classical navigation, out of necessity at first, later by passion.

Dear Reader, we wish you fair winds, a sharp horizon and a safe journey.

2020
Roden, Siebren van der Werf
Leeuwarden, Dick Huges

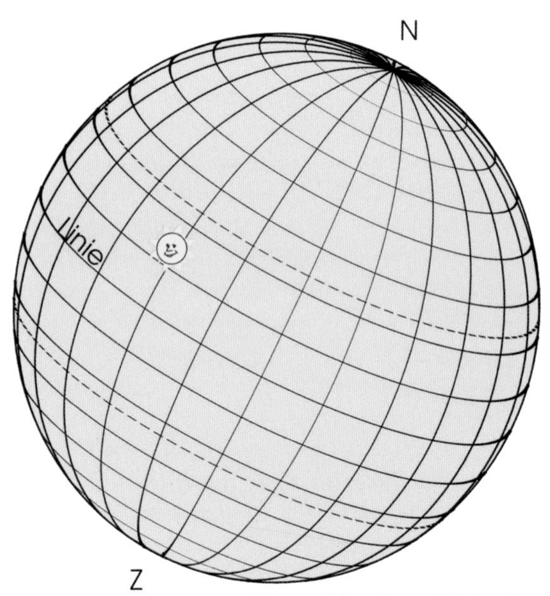

On the globe meridians (circles of equal longitude) and parallels (circles of equal latitude) are indicated. The latitude directly underneath the sun is its declination. In the time of one year, this "footprint" of the sun walks back and forth between the Tropic of Cancer, +23°.5 (North), and the Tropic of Capricorn, -23°.5 (South). More precisely: Nowadays the maximum declination is 23°26´. In the figure, both tropics are drawn as dashed lines. In Dutch and other West-European countries the equator was known as the "Linie".

History of celestial navigation

The center of the universe is at 6366 kilometers below our keel: it is the center of the Earth. Around us revolve the sun, the planets and the stars along the heavens and you can find their places at any time from nautical tables. Such was the geocentric universe for the sixteenth century navigator: the modern yachtsman, although he has learned differently in school, will have no problem in viewing the world this way.

By the end of the fifteenth century the West-European seafaring nations began sailing the oceans, initially in search of alternative routes to Asia and so to have a more direct access to its highly valued products such as silk and spices. Crossing the equator, they lost the trusted Pole Star. It sank below the horizon. In the southern hemisphere they had to rely on the sun.

The fifteenth century also saw the development of book printing. By the end of the century the works of the great astronomers had appeared in print. Two of those astronomers were most influential for developments in navigation: Abraham Zacut (1452 - ca. 1515) and Johannes Regiomontanus (1436 - 1476).

Aside from his work and teaching at the University of Salamanca, Zacut also was a rabbi of the Jewish community. His major work is the *Almanach Perpetuum,* the eternal almanac, which for the sun had astronomical tables for the years 1473 - 1476. It also gave a simple instruction for re-using these tables in later years and there was a separate table for converting the solar places along the ecliptic to declinations.

When in 1492 the Jews were exiled from Spain, Zacut was welcomed at the Portuguese royal court. Not for long, because only a few years later

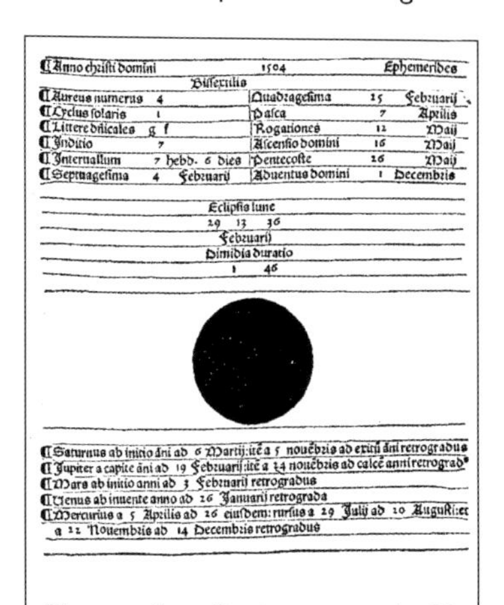

Anno christi domini 1504 Ephemerides

Bissextilis

Aureus numerus	4	Quadragesima	15	Februarij
Cyclus solaris	1	Pasca	7	Aprilis
Littere dñicales	g f	Rogationes	12	Maij
Inditio	7	Ascensio domini	16	Maij
Intervallum	7 hebd. 6 dies	Pentecoste	26	Maij
Septuagesima	4 Februarij	Aduentus domini	1	Decembris

Eclipsis lune

29 13 36

Februarij

Dimidia duratio

1 46

Saturnus ab initio ãni ad 6 Martij: itẽ a 5 nouẽbris ad exitũ ãni retrogradus

Jupiter a capite ãni ad 19 Februarij: itẽ a 24 nouẽbris ad calcẽ anni retrograd⁹

Mars ab initio anni ad 3 Februarij retrogradus

Venus ab inuente anno ad 26 Januarij retrograda

Mercurius a 5 Aprilis ad 26 eiusdem: rursus a 29 Julij ad 10 Augusti: et a 22 Nouembris ad 14 Decembris retrogradus

The page from Regiomontanus' tables for 1504, by which Columbus learned there would be a total lunar eclipse.

also Portugal made its Jewish citizens choose either to convert to Catholicism or to leave the country. Zacut's short stay had nevertheless been long enough to allow his Almanach to be translated from Hebrew into Latin. After leaving Lisbon, Zacut lived in Tunis for a number of years. He died in Turkey or in Syria.

Regiomontanus (Johann Müller) from Königsberg, Bayern, only reached the age forty. Yet he was already a legend in his life-time. His legacy is enormous. Important for navigation were his astronomical tables, pre-computed for the years 1474 - 1506. It is certain that Columbus (1451-1506) carried them on his voyages. In 1504, while stuck on Jamaica for repairs, some men of his crew had misbehaved toward the locals and their chief had ordered a stop on provisions. For 29 February a total lunar eclipse was prominently announced in Regiomontanus' tables and Columbus used this to bluff his way out of the awkward situation. He told the chief that his God was angry and would take the moon away. But he, Columbus, was willing to throw in a good word if they cooperated.

The actual computation of declinations from these astronomical tables took quite some work, looking up and interpolating in no fewer than three separate tables. The next step was therefore a logical one: don't burden the navigator with this, but instead provide him with pre-computed declination tables. And so, in the early sixteenth century the first four-year declination tables appeared in print. The choice of four years is convenient, because it contains a leap day. You can safely use such declinations again, four, eight etc. years later, a multiple of four. Over the time of one century the difference is only three quarters of a day.

The earliest navigational handbooks were Portuguese and they had splendid titles such as *Livro de Marinharia*. Best known became the *Regiment of Évora* that had tables for 1517 - 1520. Almost a century later William Barentsz (1550-1597) still used them! Pedro de Medina's *Arte de nauegar* (1545) is of later date and in its different translations it became the most widely used guide of the sixteenth century.

The Portuguese and Spanish declination tables were made on the basis of Zacut's recipe as he had formulated in his *Almanach Perpetuum*. English and Dutch tables appeared only in the second half of the sixteenth century. Widely used was the *Regiment of the Sea* by William Bourne, later continued by Thomas Hood. In the meantime the revolutionary book by Nicolaus Copernicus (1473-1543) had been published (1543): *De revolutionibus orbium coelestium* – on the orbits of the heavenly bodies - and those astronomers who embraced the heliocentric universe would base their astronomical tables on the new computational algorithms that came with it. Until then, all astronomers had used procedures that dated back to the time of Ptolemy (ca. 100-180). Halfway into the thirteenth century these rules had been laid down in the *Alfonsine*

Tables, so named after King Alfonso X "The Wise" (1221–1284) of Castile and Leon, on whose instigation the works of Ptolemy had been translated and documented.

One of the Copernican astronomers was Johannes Stadius (Jan van Ostaeyen) from Flanders. He was extremely productive and produced yearly astronomical tables for 1554 - 1606. The early English four-year declination tables were

't Eerste Deel vande

Spieghel der Zeevaerdt, vande navigatie
der Westersche Zee, Innehoudende alle de Custe
va Vranckrijck Spaignen en t'principaelste
deel van Engelandt, in diversche Zee Caerte
begrepe met den gebruijcke van dien, nu met
grooter naersticheijt bij ee vergadert en ghe-
practiseert. Door Lucas Iansz Waghenaer
Piloot ofte Stuijrman Residerende Inde
vermaerde Zeesta dt Enchuijsen.
Cum Privilegio ad decennium.

Ghedruct tot Leyden by Christoffel Plantijn/
voor Lucas Jansz/ Waghenaer van Enckhuysen.
Anno M. D. LXXXV.

Title page of Lucas Jansz Waghenaer's Spieghel der Zeevaerdt of 1585.

all based on Stadius' work.

There were quite a few changes. In particular, the maximum declination had now been fixed at 23° 28′. Zacut had used 23° 33′, which in the meantime was considered a bit outdated. Regiomontanus had chosen 23° 30′ and the same value had been adopted in the nautical handbooks of Pedro Nunez (1537) and Martín Cortes (1551). Evidently William Bourne did not yet have a proper table for converting the places of the sun to declinations. What he did: he used the table for a maximum of 23° 30′, that he knew well from the English translation of Martín Cortes' *Breve Compendio de la sphera y de arte de nauegar.* Around midsummer and midwinter he "massaged" the results so they would not exceed the "new" maximum of 23° 28′. Later, Thomas Hood would use the correct conversion.

Graetboecxken nae den
Nieuwen Stijl:
Waer in datmen tot allen Tijden vinden mach de hoochde ende leechde der sonnen / oft in wat graet de sonne alle dagen gaet: dit op de calculatie vanden nieuwen Almanach.
Noch is hier by gheuoecht van sommighe sterren hoe sy in haer linie gaen/en vande ordinantie vande noortsterre.
Item hier achter is by gheuoecht de tafel vande water ghetyden om tot wat plaetsen men is/de ghetyden van alle havenen en stroomen tot allen tyden te moghen rekenen.
Ghedruckt tot Amstelredam by Cornelis Claesz. opt water int Schrijfboeck.

Title page of the Graetboecxken nae den Nieuwen Stijl. This is a scan of possibly the only surviving copy. It is in the collection of the library of the University of Amsterdam, Special Collections, inventory number OTM: OK 61-141 (3).

Spieghel der Zeevaerdt of Lucas Jansz. Waghenaer was the first to give declination tables in the Gregorian "New Style" calendar, which had been introduced in 1582. These tables had been computed for the years 1585-1588 by Adriaen Antonisz Metius, land surveyor and later mayor of Alkmaar. Also here, the maximum declination is the new value of 23° 28′. A closer look reveals, however, that they were not based on Copernican style astronomical tables, but that instead they are Alfonsine. The English translation *The Mariners Mirrour* (1588) had the same declination tables as Waghenaer's *Spieghel,* but in the Old Style Julian calendar.

The new declination tables of Waghenaer/ Antonisz were from 1587 onward also published in a pocketsize booklet, *Graetboecxken nae den Nieuwen Stijl,* by the printer-publisher Cornelis Claesz in Amsterdam. Not everybody was happy with it. Mariners who had an astrolabe of their own, oftentimes had the calendar and the corresponding declinations engraved on the back of the instrument. And now some prophets of the New Light would tell them that this correspondence was off by ten days, because that was by how much the old and the new style calendars differed. Cornelis Claesz felt pressed to oblige his customers by publishing another booklet, after the old calendar again, *Graetboeck nae den Ouden Stijl,* 1595. He did not do so wholeheartedly: maybe to make those who had complained feel that while time had progressed they had not, he gave them the oldest declination tables available, those from the Évora regiment, which had been made for 1517 - 1520.

The turn of the sixteenth to the seventeenth century marks the beginning of a new period in astronomy and with it, for declination tables in nautical handbooks. On Hven, a small island in the Øresund, just north of Copenhagen, the Danish nobleman Tycho Brahe had studied the heavens for twenty years with an unprecedented accuracy. His work was continued by Johannes Kepler, the founding father of modern astronomy.
The renowned Dutch cartographer and maker of celestial globes, William Jansz. Blaeu worked on Hven as Brahe's assistant during the winter of 1595 - 1596. The declination tables, which he gives in his *Licht der Zeevaart* (1608) and in its translation *Light of Navigation* (1612) are based on Tycho Brahe's early solar tables for 1572 and 1573. To re-use them in later years you would have to shift them by about three quarters of an hour every four years. From 1572 to 1608, that would be thirty six years in Blaeu's case. His declination tables were computed with great care: because of this accuracy, it becomes clear that Blaeu shifted Tycho's tables over only twelve years, hence from the time of his stay on Hven. The likely conclusion is that he thought them recent at the time. Evidently he copied more than only Tycho's new star catalog.

Tycho Brahe in the garden of his former castle Uraniborg on Hven. Photo: SvdW

Prague 2018.

Photo: Ysbrand van der Werf

Our Solar System

We orbit around the sun, one full circle (or rather: ellipse) in just under 365¼ days. At the same time we rotate about our own axis and one go-around takes a day. Our moon revolves around us, a full circle in one month.
Because we go around the sun, we see it move against the background of the fixed stars, just as when you would slowly walk around a table that is in the middle of your room. As you walk, you keep looking at the table and at the wallpaper behind it. In one go-around you get to see all of the wallpaper. In the same way we see the sun make a complete circle along the skies in one year. At the end of December we see the beautiful constellation Orion in the south at around mid-night. A month later at the same hour it has moved thirty degrees to the west and by the end of February sixty. In summer we don't see Orion because it is in the direction of the sun. But next winter it is there again and by the end of December it is back in the south at mid-night. In 365 days we see the stars go round 366 times. The sun loses to the stars a full circle in a year.

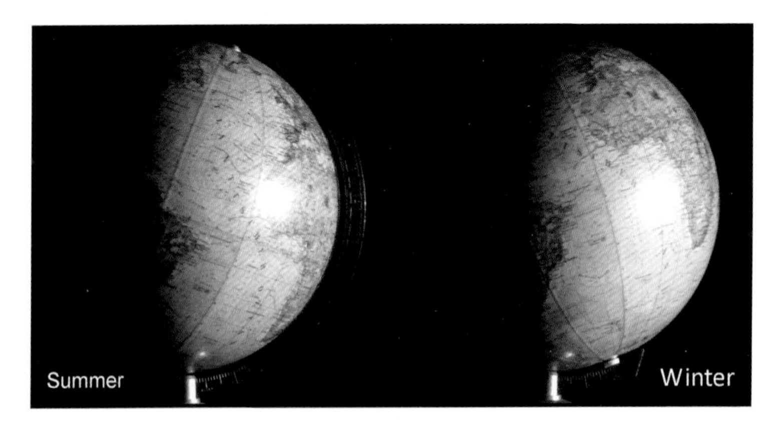

Illustrating the effect of the Earth's axis obliquity on the length of daylight.
Photo: SvdW

Our rotation axis – the north-south axis of Earth - is not perpendicular to the plane of our orbit around the sun, but makes an angle with it of close to 23°.5. Therefore the northern hemisphere gets more sunlight in the summer and the southern hemisphere in winter.
As indicated above, the sun's declination is just the latitude from where it is seen directly overhead, in the zenith. Midwinter, 21 December, it is –23° 26′ (minus = S) and the sun is above the Tropic of Capricorn. Midsummer, 21 June, is the longest day for those who live in the northern hemisphere. The sun is then to the north side of the equator above the Tropic of Cancer, at 23° 26′ N. On 21 March and 23 September the sun crosses the equator, in March from south to north, in September from north to south. Day and night are then equally long everywhere on earth and for this reason these equator crossings are called equinoxes. As a further matter of curiosity, during equinox days all shadows trace paths from west to east along straight lines over the ground.

Meridian altitude and zenith angle

At meridian transit you always see the sun in the south if you are north of the Tropic of Cancer. But if you are south of the parallel where the sun stands directly above, then you see the meridian sun in the north. The figures below illustrate the two situations. In both cases the zenith angle, also known as "zenith distance", *ZEN*, is the angle by which the sun is "out of plumb", and it gives you directly the difference between your latitude and the declination of the sun. *ZEN* is always counted positive.

LAT and *DEC* are the usual abbreviations for latitude and declination respectively. Both can be either positive or negative. The declination tables at the end of the book (pp. 71-87) give the absolute value and distinguish between positive and negative by different colors: red for positive (North) and blue for negative (South).

An observer who sees the meridian sun in the south finds himself to the north of the sun. He finds his latitude, *LAT*, by adding the zenith distance to the declination, because the latter is just the latitude of the sun's footprint, and *ZEN* is the difference between the two latitudes:

LAT* = *DEC* + *ZEN

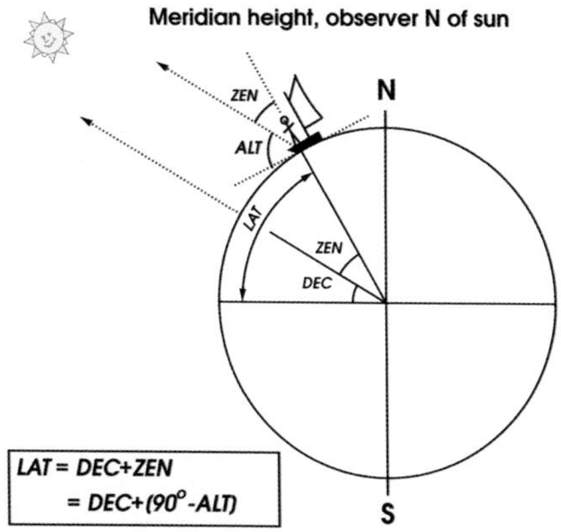

When the meridian sun is to the north, *ZEN* is still the difference between the latitudes of the sun and that of the observer, but now he must subtract *ZEN* from *DEC*:

LAT* = *DEC* - *ZEN

Meridian height, observer N of sun

ZEN
ALT
N
LAT
ZEN
DEC
S

LAT = DEC+ZEN
= DEC+(90°-ALT)

We will come back to this later, but first something must be said about the way the zenith distance is usually found in practice. Both illustrations indicate that it is the complement of the altitude, *ALT*, which is the angular distance above the "true horizon". This true, or astronomical, horizon, or "level" as a carpenter would call it, is at right angles to the zenith. Thus: ***ZEN* = 90° - *ALT***

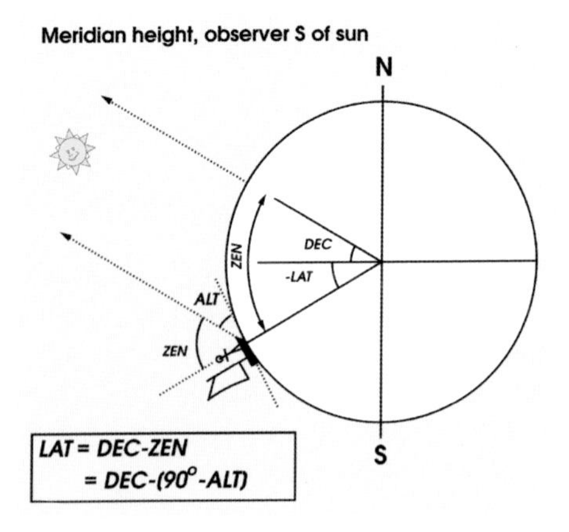

When the height of the sun is taken with a sextant, then it will be measured from what is seen as the horizon. Because the observer himself is a few meters above the water, he sees the horizon just a bit lower than level. The difference is the horizon dip, or dip for short. A second correction must be made because light rays are not perfectly straight and atmospheric refraction bends them downward. As a result the sun and the stars are seen just a bit higher up in the sky than they are in reality.

Measuring height with a sextant

The idea to use mirrors for finding the height of the sun should probably be credited to Hooke, and be dated at around 1666. The first practical instrument based on this idea is due to Hadley, who introduced it in 1731. Independently, also the American optician Thomas Godfrey designed a double-reflecting octant in the same year. The principle is shown in the accompanying figure, which we took from *Wrinkles in Practical Navigation*, by Captain S.T.S. Lecky, edition 1920, who in turn took it from Herschel's book *Astronomy*.

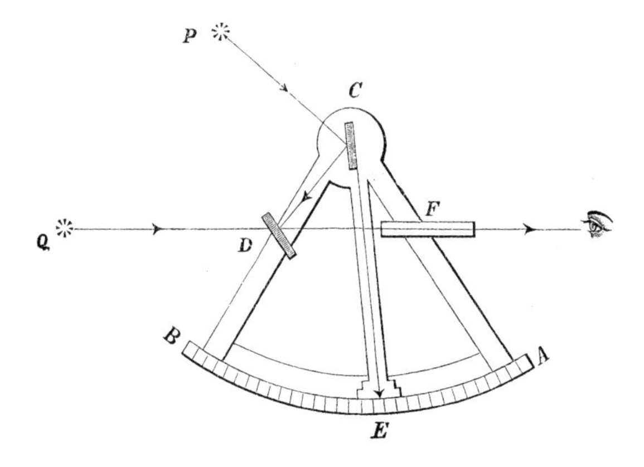

Light emitted by a celestial body **P** is reflected by the index mirror **C** and a second time by the horizon mirror **D**, which is fixed on the frame of the sextant. After reflection in the horizon mirror the light reaches the eye of the observer through a telescope, **F**. The index mirror is mounted on the index arm that is movable along the graduated scale, from which the angle is read, in the figure in **E**.

For a sextant this graduated arc, ***AB***, covers 60°, one sixth of a full circle, whence its name. Changing the reading angle of the arm, and therefore of the index mirror, along the graduated arc, results in precisely twice this change for the angle ***PCD.*** By this doubling a sextant covers an angular range of 0° - 120°, usually just a bit more.

Hadley's first instrument was not a sextant, but an octant. The graduated arc covered one eighth of a circle (45°) and thus allowed for measuring angles up to 90°. This has led to some confusion in literature, where sometimes it has been called a quadrant. But the name derives from the length of the arc ***AB***, not from the double angular range that can be measured with it.

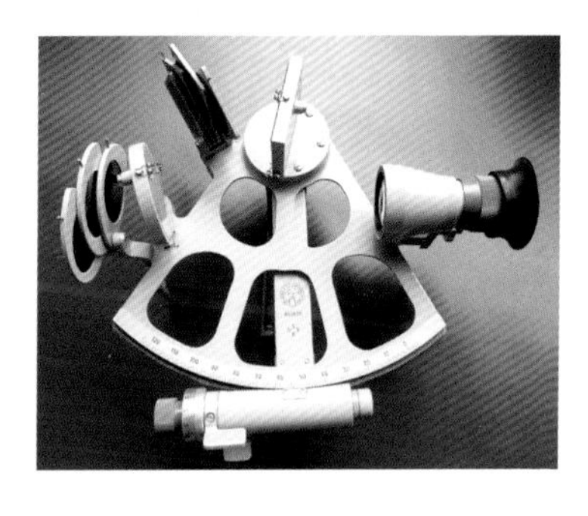

A modern sextant with worm-wheel drum dial and full-sight mirror.
Photo: SvdW

In the traditional horizon mirror the right half is silvered to reflect the light by which the heavenly body is seen. The left half is fully transparent, giving the observer a direct view of the horizon, **Q**. Nowadays also a half-silvered horizon mirror is a convenient option. In such a full-horizon, or full-sight, mirror both the direct view of the horizon and the reflected image of the sun are seen simultaneously. By adjusting the reading of the sextant to make the reflected image of a celestial body coincide with the horizon, its height is found. For the sun, usually its lower limb is brought down to the horizon. This guarantees a sharper reading.

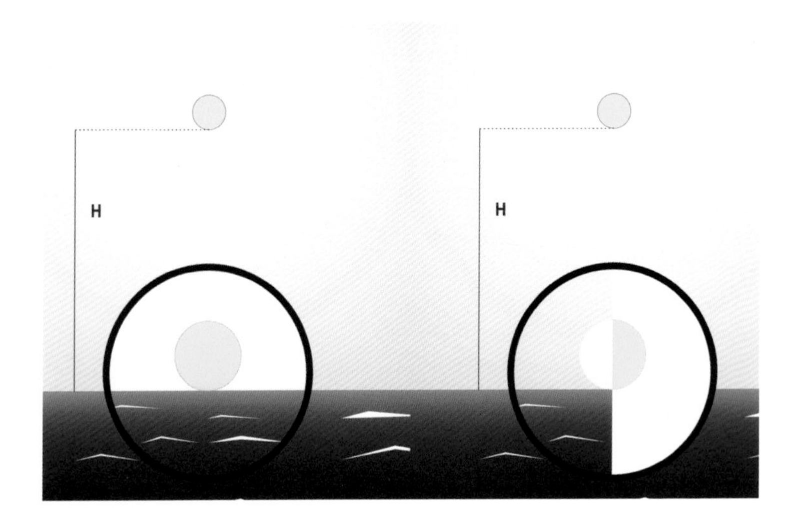

A lower limb observation. Left: seen through a full-sight mirror. Right: with a traditional half-sight mirror. Note that the image of the sun can also be distinguished in the left-hand part, because a flat glass plate still reflects a small percentage of the light intensity.

Half the diameter of the sun may be looked up from the daily pages of the *Nautical Almanac*. If you don't have an almanac, it will usually be sufficiently accurate to adopt the yearly average, which is 16′. Since we want the height of the sun's center, this semidiameter must be added to the sextant reading.
You may also take sights without the telescope. Your field of view is larger, which makes finding the sun easier. This is a definite advantage when looking for the right moment to ride a wave that will give you a good horizon. The little loss of accuracy, because the image is not magnified, is small and acceptable. Look with both eyes open. With your right eye you look through the horizon mirror, with your left eye directly.
Bernard Moitessier used this method during his double solo circumnavigation in 1968-1969. It is described in his book *La longue route* (*The long way*, UK).

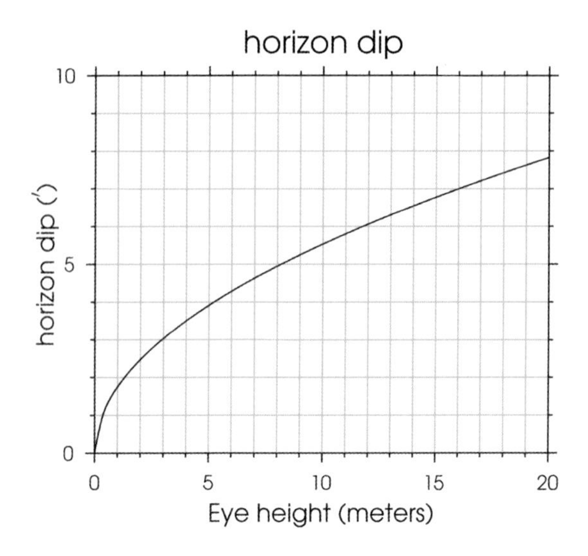

Formula: dip = 1′.75 √h, where h is the eye height in meters.
The distance of the horizon is 2′.1 √h nautical miles.

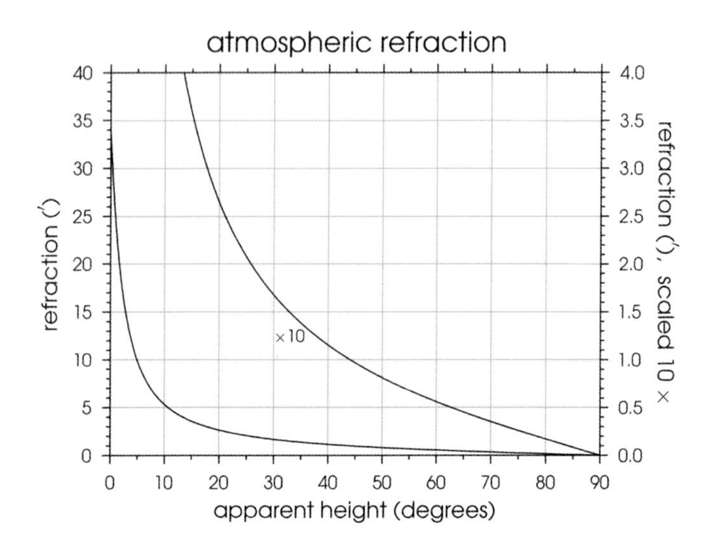

Using a pocket calculator: for apparent heights (H_A) larger than 10°, the formula refraction = 1′.0/tan(H_A) is sufficiently accurate.

Refraction and horizon dip.

Light rays are bent slightly downward in the atmosphere. This effect is called refraction. The result is that we see a celestial body a bit higher than it is in reality. But also we see the horizon just a bit higher than it would have been seen without refraction.

This bending effect is strongest for rays by which we see light from close to the horizon. This causes especially horizon dip to be somewhat variable and dependent on the temperature profile of the air in the first 100 m above the sea.

A sextant reading, *H* (height), must be corrected for both dip and refraction. We give these corrections in the form of two graphs, both computed for the same standard atmosphere that is also used by the *Nautical Almanac*. To do it correctly, the dip correction should be applied first, giving the apparent altitude, H_A, which is now measured, not from the apparent horizon but from "level", the true or astronomical horizon. Refraction tables are computed for apparent altitudes as entry. Subtracting the thus found refraction from H_A gives the true altitude, which we shall denote by *ALT.*

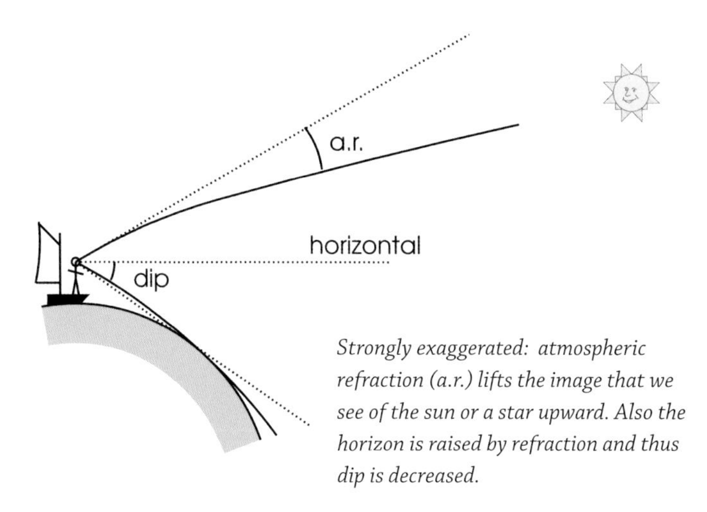

Strongly exaggerated: atmospheric refraction (a.r.) lifts the image that we see of the sun or a star upward. Also the horizon is raised by refraction and thus dip is decreased.

The Earth is round.

Summarizing:

sextant reading	H	45° 21′	
horizon dip	dip	2′.8	minus
refraction	a.r.	1′ 0	minus
semi-diameter	s.d.	16′.0	plus
True altitude	ALT	45° 33′.2	

Example: for a sextant reading of 45° 21′ of the sun's lower limb and an eye height of 2.5 meter, the true altitude of the sun's center is found as worked out in the box.

A final remark. A yachtman who takes the height of the sun will want to get as good as possible a view of the horizon and when there are waves he will try to make his sighting from the top of one. There exists a persistent misunderstanding among yachtsmen that you should then add half a wave height to your normal eye height before entering the dip table. However, what you see as the horizon is made up of all the wave tops that in the distance cannot be distinguished individually. And of them a good percentage will be higher yet than the one from which you took your sight. If you want to do it right, you will have to *subtract* a fraction of the estimated wave height. Twenty percent of the average wave height is what comes closest. The correction, however, is minor and no man is overboard if you ignore it.

The regiment of the North Star

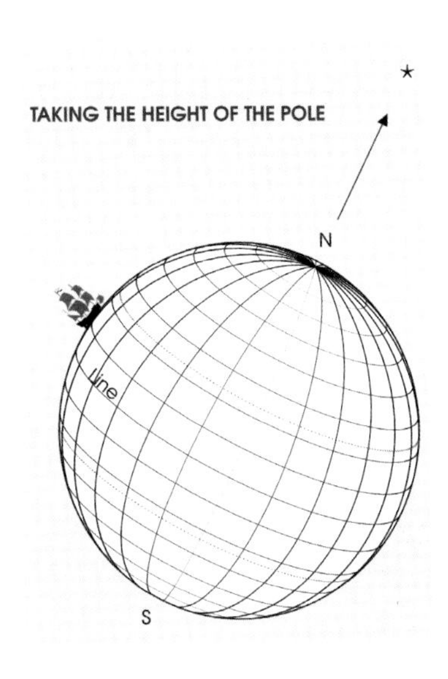

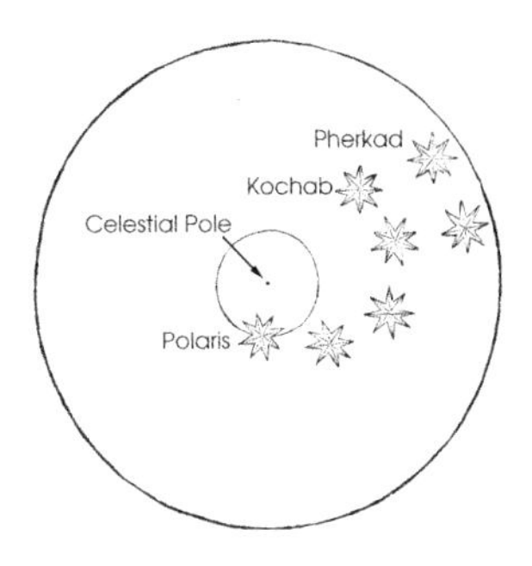

From Pedro de Medina's book Arte de Nauegar (1545), explaining how at the time Polaris circled around the celestial north pole and how the "Guards", Kochab and Pherkad served as an hour hand.

The oldest method to find your latitude at sea is by taking the height of Polaris, the Pole Star or North Star, the brightest star in the constellation Ursa Minor. In the illustration we see a seventeenth century ship that has just crossed the equator.William van der Decken, the Flying Dutchman on his way home. Once every seven years he is allowed on land. If he then finds a woman who will agree to share her life with him, his curse will be ended. If not, then he must go to sea for another seven years.

The Pole Star comes in sight and with every degree he advances north, the Pole Star rises higher above the horizon by one more degree. Indeed: height of the pole = latitude.At least… almost.The place of the true north is the celestial pole, a fictitious point in the starry skies on the extension of the Earth's rotation axis.The Pole Star revolves around this celestial north pole on a tiny circle of only 38′ in angular distance. In olden days, around 1500, that distance was larger, about three and a half degrees. Nautical handbooks would give methods to find the true celestial north pole.

In the sixteenth century the place of the true north was found at about a quarter of the distance from Polaris to the second and third brightest stars in Ursa Minor, Kochab and Pherkad, the "Guards". They are indicated below on the *Height of the pole correction disc.* The circle on which Polaris travels around the celestial pole is too small to recognize with the naked eye where exactly it is relative to the true north. By using the Guards, however, you can distinguish it easily and the line between them and Polaris serves as the hour hand of a celestial clock. The sixteenth century "Regiment of the North Star" was a table that for different orientations of the Guards gave the correction that you had to add to or subtract from a measured height of Polaris to find the true height of the pole, which then at the same time was your latitude. Also the stars Dubhe and Merak in Ursa Major, the "Pointers", could be used in a similar way.

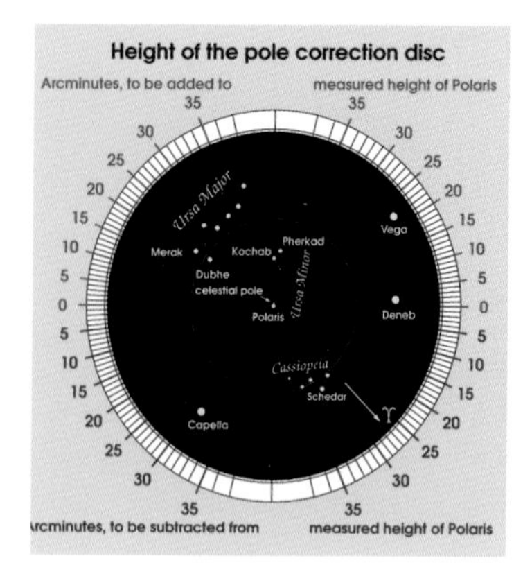

The modern pole height correction disc.

Today Polaris is much closer to the celestial pole and its circle around it is correspondingly smaller. The pole height correction disc shows the northern sky. In the time of one day all stars make one counter-clockwise rotation around the celestial pole. The correction disc (also on the back cover of this book) is the modern equivalent of the regiment of the North Star. Ursa Minor and Ursa Major are shown on the disc and also Cassiopeia, easily recognizable by its characteristic W-shape. Also three very bright stars are indicated: Vega, Deneb and Capella. They are farther off from the pole, but are easily recognized.

To find the correction, turn the disc so that it corresponds to how you see the sky in reality. Then read off from the top (12 o'clock) how much you should add or subtract from your measured height of Polaris (corrected for dip and refraction) to find your true height of the pole, your latitude. As an example, see the illustration on p. 43 which shows the disc rotated to match the situation as it was seen in reality.

There is another, more indirect method that is described in modern nautical almanacs. The idea is this: all stars go around the celestial pole equally fast, about four minutes faster than the sun in one go-around. Thus all their angular distances remain constant. The almanacs therefore do not give daily tables for

Polaris has not always been useful as a "pole star". In the second century AD, Ptolemy established its distance to the celestial pole as about 12°. By the end of the fifteenth century the distance had decreased to three and a half degrees. Today, Polaris is off from the celestial pole by only 38´ and its distance is still decreasing. In the year 2100 Polaris and the pole will pass each other at a distance of 28´.1. Thereafter they will gradually move away from each other. About ten thousand years from now, Deneb, the brightest star in the constellation Cygnus, will be the new pole star.

each star separately, but only for the fictitious point where the sun crosses the celestial equator from south to north on 20 or 21 March. The point is called Aries, the vernal equinox, where the sun enters the zodiac sign of the same name. The symbol for Aries is ♈, symbolizing the horns of a ram.
Greenwich hour angles for Aries are given in the daily pages and for each of the selected fixed stars the hour angle difference with Aries is given: the *sidereal hour angle* or *SHA*. This makes it easy to find the Greenwich transit time for any star.
The direction of Aries is indicated on our pole height correction disc. Nowadays, in 2021, Polaris is close to three hours behind Aries. To be precise: beginning 2021 their difference in *sidereal hour angle* is 315° 15´, which, subtracted from 360° makes 44° 45´, indeed just under three hours in time. Hence, if you know where Aries is, you immediately find the hour angle of Polaris and its place on the tiny circle on which it revolves around the celestial pole. With this method Aries has taken over the role of the Guards. But the pole height correction disc is much easier to use and it is accurate enough. The disc shown here and on the back cover of the book has been computed for 2023, when the distance of Polaris will be 38´. You can perfectly well use it in later years, because over the next 10 years the distance to the pole will gradually decrease by only 2´. It will be 36´ in 2033.

Atlantic Ocean 1984. Skipper (DH) takes meridian altitude.

Photo: John Stuart

Finding your latitude from a meridian altitude

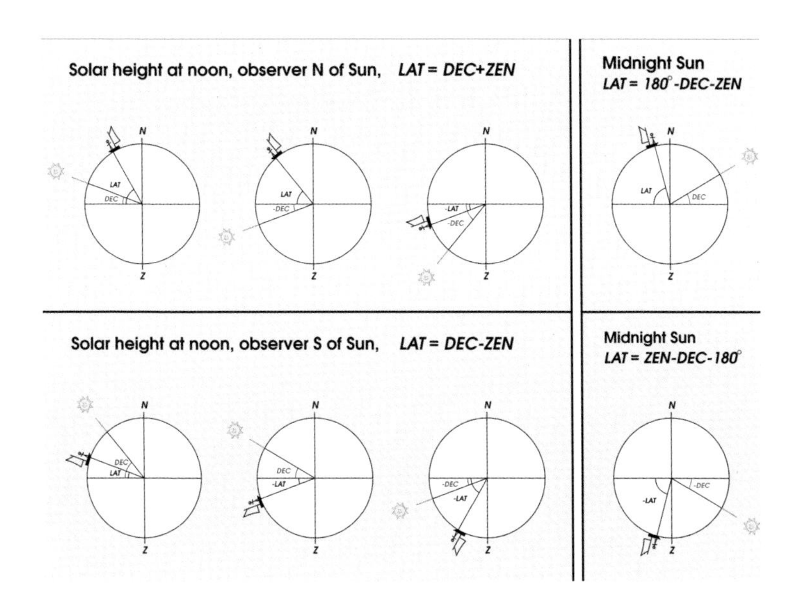

The illustration above shows the different possible scenarios, when taking a meridian altitude. We have already shown that zenith distance and altitude are complementary:

ZEN* = 90° - *ALT

ALT is the true altitude, corrected for refraction and horizon dip and for the sun's semidiameter in the case of a lower- or upper limb observation. If the sun is south of you, then you are to the north of it and you find your latitude by adding *ZEN* and *DEC*. This is true for northern and southern declinations alike. If the declination is south, then just take along its minus sign.

LAT* = *DEC* + *ZEN

If you see the mid-day sun to the north of you, then you subtract the zenith angle from the declination.

LAT* = *DEC* - *ZEN

An example: a man, living in Harlingen, The Netherlands, finds that at midsummer the sun rises to 60° 16′ above his true horizon and at midwinter to only

13° 24′. In the first case the declination is 23° 26′, in the latter -23° 26′. Working out his latitude for both cases he finds:

Midsummer: $LAT = 23° 26′ + (90° - 60° 16′) = 53° 10′$ N
Midwinter: $LAT = -23° 26′ + (90° - 13° 24′) = 53° 10′$ N

His brother lives in Melbourne, Australia. At midsummer he finds that the sun rises to 28° 46′ and at midwinter to 75° 38′. In both cases he sees the sun in the north. He finds his latitude as follows:

Midsummer $LAT = 23° 26′ - (90° - 28° 46′) = -37° 48′ = 37° 48′$ S
Midwinter: $LAT = -23° 26′ - (90° - 75° 38′) = -37° 48′ = 37° 48′$ S

Then the midnight sun: the right-hand panels in the above illustration show how latitude is found in that case.
On 23 June 1596 Gerrit de Veer described how William Barentsz and his men, anchored at the north-west side of Spitsbergen, took the height of the midnight sun. From his diary it becomes clear that it was the night of 22-23 June: *.... and found it to be elevated above the horizon 13 degrees and 10 minutes, his declination being 23 degrees and 28 minutes; which subtracted from the height aforesaid* [he means it the other way around: the height is subtracted from the declination], *resteth 10 degrees and 18 minutes, which being subtracted from 90 degrees, then the height of the Pole there was 79 degrees and 42 minutes.*
[Translation by William Phillip, 1609].

Here, De Veer uses the scheme, given in the top-right panel of the illustration:

$$LAT = 180° - DEC - ZEN = 180° - DEC - (90° - H) = 90° - (DEC - H)$$

On a historical note: Barentsz had, no doubt, several declination tables with him.
His own *Caertboeck der Midtlandsche Zee* (1595) had the old Évora tables. The more recent Dutch translation of De Medina's *Arte de nauegar* (1580) had been left behind in their winter shelter and was found there when the remains of their house were discovered in 1871. This declination of 23° 28′ seems to have been taken from the *Graetboecxken nae den Nieuwen Stijl.*

Longitude and latitude from sunrise and sunset

From the times of sunrise or sunset you can find your longitude with reasonable accuracy. Tables for the periods of daylight are given as appendices to this book (pp. 51-69), for all latitudes and all declinations.

The way it works is best explained by an example. Suppose the date is 2 August 2023 and you are underway from the east coast of the United States to the Azores. You watch the sun set and you time its last light at 22:58 GMT. By your last meridian altitude and by dead-reckoning your latitude is close to 40° N. From the appendices that list the declinations, you find that it is 17° 45′. If you are a perfectionist you may interpolate in the declination tables, since noon at Greenwich, for which the tables are given, is already 11 hours ago. You will then adopt a declination of 17° 39′. By a further simple interpolation in the day length tables you find that for the setting of the sun's upper limb the period of day length is 14 hours and 14 minutes. Thus, *half a day length, 7:07 hours, ago the sun must have made its transit through the meridian where you are now, at this very moment.* That was at 15:51 GMT.

The graph of the Greenwich transit time tells you that the sun crossed the 0° meridian at 12:06, hence earlier than for you by 3:45. The sun moves west at a speed of 15 degrees in one hour and you conclude that your longitude is 3×15°+45×15′ = 56° 15′ west of Greenwich. In your chart you put a cross at 40° N, 56° 15′ W.

If you are not quite sure about your latitude, then you may do the same exercise once more, for instance assuming 39° N. Then half the day length is 7:05, as it follows from interpolating in the tables: two minutes shorter than it is for 40°. In these two minutes the sun does half a degree in longitude and you must shift your longitude to the west by that same amount. In the chart you put a second cross at 39° N, 56° 45′ W. The line through both crosses will be your position line.

Instead of making a second calculation, you may of course also take the compass bearing of the sun and draw your position line through the first cross at right angles with the bearing of the setting sun.

Suppose now that you did not only time the sunset, but also the sunrise and suppose further that you found a day length of 14:17, three minutes more than you had looked up for 40°. In the same way you then find that for 41° N the day length is 14:19. What you measured is in between and interpolation suggests 40° 36′ as the best guess for your latitude.

Finding your latitude from sunset or sunrise works best if the declination is large. Around the equinoxes the period of daylight is the same everywhere. Finding your longitude will of course still work.

We give the day length tables in two versions, both for the lower limb of the sun on the horizon and for the setting of the upper limb. Which one you choose is up to your own preference.

How accurate is all this? The day length tables have been calculated for the same standard atmosphere that has also been adopted by the *Nautical Almanac*. That is: for a pressure of 1013.25 hPa and a temperature of 10 °C at sea level, slowly decreasing with height by 0.0065 °C per meter. The same atmosphere has also been used for computing the horizon dip that is implemented in the day length tables. Horizon dip is variable and depends on the temperature profile over the first tens of meters above sea level. In the example that we use above, the day length for the upper limb is 6 minutes more than that for the lower limb. In half that difference, 3 minutes, the sun sinks by 32′, its diameter. Hence, about 10′ in a minute. And if the horizon dip is off by 10′ from its normal value, the setting sun will be noticeably deformed, because you deal with a significant mirage. And even then: in this one minute

Sunset on 24 July 2014, seen at 19:41 GMT from Stavoren, NL, at 52° 52′ 51″ N, 5° 21′ 32″ E. Photo: DH

As an aside: note the flattening of the sun. Refraction lifts the lower limb by 5′ more than the upper limb. If you don't see it immediately, give this page a quarter turn.

the sun moves by 15′ from East to West. Because we have also rounded our day lengths to the nearest minute, this estimate of 15′ will at the same time be the typical accuracy of the method.

To show that it really works quite accurately, we give an example of a sunset, this time for the lower limb on the horizon and photographed from a known location. We adopt the known latitude and we want to verify the longitude. Reading the declination from the tables for 2022 (2014+8 years) and interpolating for the about 8 hours that have passed since noon at Greenwich, we find that the declination was 19° 45′. The latitude is 52° 52′.8 N. Interpolating the day length table for both declination and latitude, we find that half the day length is 15:52:30/2 = 7:56:15, which subtracted from the time the photo was shot, gives a local transit at 11:44:45. Transit at Greenwich was at 12:06:30 GMT, a difference of 21m 45s. This results in a longitude of 5° 26′ E, only 5′ more east than the actual location or, at this latitude, a difference of 3 miles, which is about the distance to the horizon as seen from the deck of a small yacht.

In a cruder form the method that we have described above is also used by biologists for reconstructing the migration routes of birds, such as the arctic tern. On their nesting locations in the North they are ringed with light-sensitive chips that record the approximate daily periods of light and dark. The chips are retrieved and read out when (and if) they return to their nests.

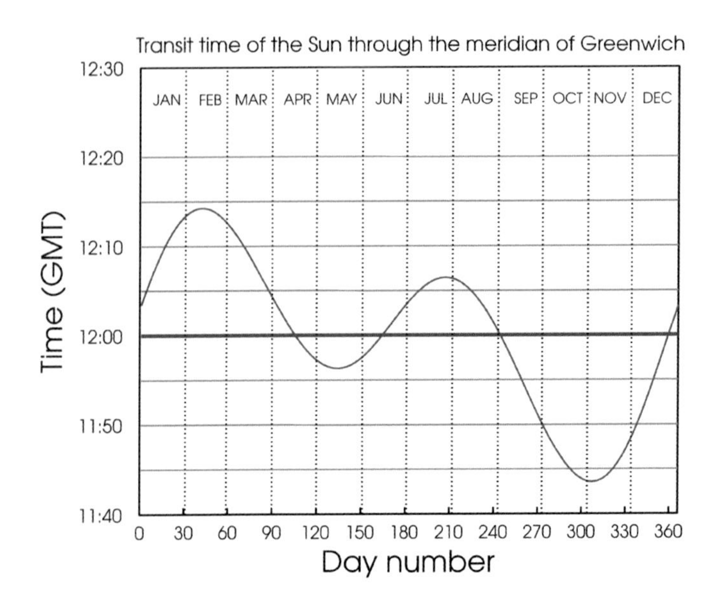

The curved red line in the above graph gives the transit time of the sun through the Greenwich meridian in GMT.
In the first months of the year the sun's transit is later than 12 GMT: the sun lags behind the clock. The difference is maximal around half of February. From September onward till around Christmas the situation is reversed and the sun is ahead of the clock with a maximum in early November. This difference, clock time minus solar time, is just the negative of the Equation of Time (EoT), which is tabulated in nautical almanacs. Its oscillating shape is caused by two factors, 1) the fact that the earth's orbit around the sun is not a circle but an ellipse, and 2) because the rotation axis of the Earth is not perpendicular to our orbit around the sun, but is inclined by 23˚.5.

Emergency astronavigation during an ocean crossing

Preface

Famous discoverers and navigators like Bartolomeu Dias (1487), Vasco da Gama (1497), Christopher Columbus (1492), Ferdinand Magellan (1519), Francis Drake (1577), William Barentsz (1596), Olivier van Noort (1598), Henry Hudson (1609) and many others, all could find their way over the oceans by the Pole Star and the Sun. The technique they used was "Sailing the Latitude". But nowadays our methods of navigation are different. We use electronics to find our position at sea. Before 1990 we had the Radio-Direction-Finder and Decca. Today, we rely on the extremely accurate GPS satellites. Just buy yourself a clever electronic device. The "electronic brain" does the intellectual work and guides you flawlessly to your destination. I (DH) considered myself lucky that I didn't need these "romantic" but also a bit frightening "old-school" navigation techniques any more.
This was my thinking when I prepared for my first Atlantic crossing. Alas, I was wrong. Electronics is really wonderful indeed ... as long as it functions.

First crossing to America

On the 2nd of June 1984 the Dutch sailing yacht *Gladys* and myself left Plymouth UK bound for Newport, Rhode Island, USA as participants in the OSTAR (Observer Singlehanded Trans Atlantic Race), a solo sailing race from England to America. A difficult "uphill" voyage because the direct route is against the predominant westerly winds and against the east-flowing Gulfstream. After four days at sea the engine, needed for generating electricity, stopped.

2nd of June 1984; Gladys at the start of the OSTAR. Photo: Jan Prakken

The cause: seawater in the cylinders. One day later the batteries were flat. So no more navigation lights, no RDF, no weather forecasts, no cabin lights during the dark hours.

What now skipper? Just head on for the US or turn round and sail back to England? Turning back meant safety. I knew the English south coast. Going on meant sailing into the unknown without the help of modern navigation conveniences. At this time, America was far away behind the horizon. It would take at least five weeks to get there. I was not afraid that I would miss that continent. But what about landfall once there? Making landfall on an unknown coast with only a sextant and without the support of the RDF, which I had bought for precisely this purpose, was frightening to say the least. Is my rudimentary knowledge of astronavigation enough to bring me exactly to that specific spot 3000 miles behind the horizon? I hesitated. What to do?

Brenton Reef Tower. Photo: DH

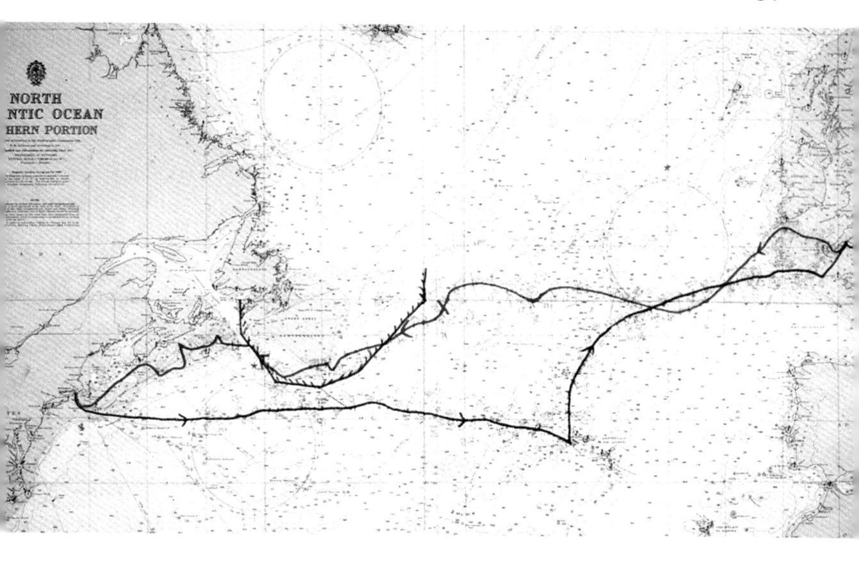

Red line: outward journey to Newport, USA. Green line: limit of 1054 icebergs. Black line: return journey to Plymouth, UK by the Azores. Photo: DH

After half an hour of contemplating the pros and the cons I decided, against logical common sense, to sail on. One thing was to my advantage. *There was time and space enough for practice.* I speculated that five weeks would give me enough time to study and master the theoretical and practical challenges of sextant navigation and I hoped that it would be sufficient to make a safe landfall.

After 35 days of contrary winds, storms and calms, cold and fog and 1054 officially registered (US coastguard) but invisible icebergs around, at last we approached America. I could even smell the land. The sweet scent of pollen. Great of course, but at the same time also the moment of landfall was coming nearer! Had my sextant sights been accurate and had my sight-reduction math been correct? All very exciting!

South of Newport we steered north and began approaching the coast. Unwelcome fog came in, visibility decreased. This complicated things a lot. I moved on. And then one hour later… I heard the deep sound of the foghorn of Brenton Reef Tower, located a few miles south of Newport. Ten minutes later its steel construction showed up right in front of *Gladys*' bow. I had a fix! What a relief, what a kick! With just a sextant, a watch, an almanac, a sight reduction table and, for me of prime importance, a "how-to-do" astro booklet, it was possible, after five weeks alone on an empty sea, to make landfall *spot on*.

Thanks to the many generations of Great Minds, who have made this possible.

From the USA to the Azores

After a few weeks of relaxing in Newport (41°.5 N) we set out to sea again. Now back to the Old World, Europe. I write here "we", because John, an Englishman, had joined me. He wanted to go back to England and I was looking for a crew. The first part of the return journey went to the Azores. A medium stretch of about 2000 miles at a course of 90 degrees (east). The not sought after, yet achieved, navigation success of the outward journey had aroused my interest in traditional astro techniques. This easterly course was therefore a great opportunity to try out the age-old technique of "Latitude Sailing".
We decided to follow the 40-degree parallel and to determine daily our ship's latitude with the help of the *North Star and the meridian height of the sun*. The Azores are at 38°.5 N, close to this course.
Columbus did the same in 1493 when he sailed from the present Haiti / Dominican Republic (20° N) via the Azores island of Santa Maria and on March 15, 1493, after a two-month journey, he reached Lisbon (38°.5 N), which lies at almost the same latitude as the Azores. A historic voyage that would change the world!

Gladys is "Sailing the Latitude" along 40° N to the Azores. Photo: DH

Latitude by the height of Pole Star

The Pole Star, proper name Polaris, is a rather faint star in the northern sky, just like so many others. But Polaris is special because it stands nearly right above the North Pole of the earth. Therefore all stars orbit around it, a full circle in 24 hours while Polaris is almost stationary. And the beauty of this is that the vertical angular height of Polaris above the ship's horizon corresponds very nearly to the ship's latitude north of the equator.

So how do you find the Pole Star and how do you measure its angular height above the horizon? Polaris is easy to find. First look for the Great Bear, also known as the Big Dipper (USA), or the Plough (UK). The official name of this constellation is Ursa Major. Everyone knows it. In the upright side of the Plough are two easily recognizable stars, Merak and Dubhe, their connecting line pointing to the Pole Star. If you extend the length of this side of the Dipper five times, you will find there a faint star. This is Polaris.

The procedure to "shoot" Polaris with my sextant underway to the Azores was as follows: during the morning twilight there is a short period of about ten minutes in which you can see both the horizon and the Pole Star. First I position the index arm of the sextant according to my dead reckoning latitude. Next the northern horizon is viewed through the telescope (magnification 4x). The flickering light that shows up, reflected by the mirrors, is Polaris. Next, I adjust the micrometer to bring Polaris exactly on the horizon and read off the vertical angular height from the index arm. Also I note the Greenwich Mean Time (= GMT) of the observation. Next I do the index correction. Now I know the sextant-measured height of Polaris above the horizon, which gives the ship's latitude… *by approximation.*

By approximation, because this sextant height must be:
1) Corrected for dip and refraction to get your true Polaris altitude (= *ALT*).
2) Corrected for the eccentricity of Polaris from the celestial north pole.
Only then you know the angular height of the celestial north pole, which equals your latitude.
1) This correction depends on a) dip, the correction for your height of eye above sea level and b) refraction which depends on the height of Polaris above the horizon. Refraction will lift Polaris a bit. On a sailing yacht the combined correction for dip and refraction is normally around minus 4′. See also the chapter on refraction and dip. pp. 23-24,
2) What about this eccentricity correction? Polaris' place in the skies does not precisely coincide with the celestial north pole. Astro-technically: in 1984 the declination of Polaris was 89°12′ N. This means that in 24 hours Polaris travels around the celestial north pole on a small circle that has a radius of only 48′, too small to recognize by the naked eye where exactly the Pole Star is, relative to the true pole. It may be below or above it or at the same altitude. Yet, an error of maximally 48′ is too large to be neglected. After all, 48 arcminutes correspond to 48 nautical miles (nearly 90 km) on your chart. So, to find the altitude of the celestial pole itself, you still have to correct Polaris' true altitude

for its eccentricity. The maximal correction of 48′ is to be added when Polaris is straight below the pole, at 6 o'clock, and should be subtracted when Polaris is right above the pole at 12 o'clock. At 3 and 9 o'clock, no correction is needed. Thus the correction is variable between -48′ and +48′.

In 1493, during the return voyage by Columbus to Spain, Polaris' eccentricity was around 3°.5. That is more than 200 sea miles or 370 km. Columbus did correct for this eccentricity by using the so called "Regiment of the North Star". The Regiment explained how the position of Polaris around the celestial north pole could be read off from its orientation relative to the so called "Guards", Pherkad and Kochab in Ursa Minor or to the so called "Pointers", Merak and Dubhe in the constellation Ursa Major. These Guards and Pointers mimic the hour hands of a celestial clock and the correction for eccentricity was tabulated in the Regiment for their various orientations in the sky.

Nowadays we use Aries as a pointer for Polaris. Aries is a fixed fictitious point on the celestial equator and hence its declination is 0°. It is used as the 0° reference meridian for longitude of all the stars on the celestial sphere, just as Greenwich is the 0° meridian for longitude on earth. The westward angular distance between Greenwich and Aries is called its Greenwich Hour Angle (*GHA*-Aries). It can be looked up from the *Nautical Almanac* for any time. Knowing *GHA*-Aries, you can determine the Local Hour Angle of Aries by using your dead reckoning longitude (see the calculation below). *LHA*-Aries is the angular distance between your local meridian and the meridian of Aries. Once you know your *LHA*-Aries, you use it as an entry in the Almanac's table of the correction for eccentricity that is to be added to, or subtracted from Polaris true altitude to find the true altitude of the celestial pole, which then equals your ship's latitude. See also the chapter "The Regiment of the North Star", pp. 25-27.
Later on (p.43) I give a detailed example of a Pole Star measurement.
Today, Polaris is trailing behind Aries by almost three hours (45°). During my crossing in 1984 this was shorter, about 2¼ hours (35°). The reason for this difference lies in the precession of the rotation axis of the earth. The celestial pole traces a full circle around a point in the sky somewhere in the constellation Draco in about 26000 years. See also the box on p. 27.

Should you find this eccentricity correction procedure too complicated or, in case you don't have an Almanac to find the *GHA* of Aries, there are a few alternatives.

1) On the back cover of this book there is the Height of the pole correction disc. This is the modern version of the "Regiment of the North Star" as used in the old days. It works like this: hold the disc so that the orientation

of the Big Dipper on the disc is the same as you see it in your night sky. Then read off the correction (in arc-minutes) at 12 o'clock on the outside ring. Mind the plus or minus sign. Further on, on p. 43, there is a figure, which demonstrates this.

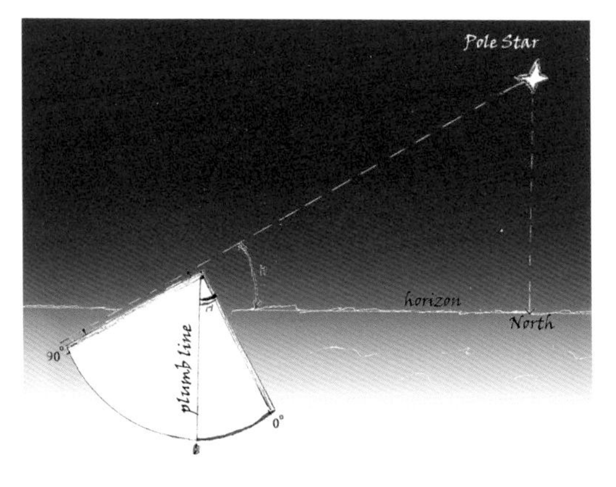

Finding the height of the Pole Star at any time of the night with an improvised self-made quadrant. See also the photo on page 46.

2) As long as you keep your latitude, for example during an Atlantic west to east crossing to the Azores, you can measure the altitude of Polaris in the evening and then again in the morning. When there is about a 12 hours' time difference between both measurements you can average both true Polaris altitudes. This averaging gives you your true north pole altitude and your true latitude because the eccentricity correction in the evening is the same as 12 hours later in the morning but reversed.

Note that there is also a practical problem, should you want to measure the height of Polaris at night. During the dark hours you can see Polaris but you don't see your horizon. And in daytime you will see your horizon but you don't see Polaris. There are, however, a few solutions to this problem.

a) During twilight in the evening, and especially in the morning, both the Pole Star and the horizon can be seen. Say about ten minutes. A short time window but still a chance. I did this during the return journey in 1984 from America to the Azores. It works fine when visibility is sufficient. The observation itself takes only a few minutes under normal sea conditions.

b) You can also create your own horizon by constructing a quadrant with a plumb line. After all, the plumb line is perpendicular to your true horizon. Such a quadrant is easy to make on board in case of an emergency. Just saw a quarter circle out of a piece of plywood that can always be found somewhere on board. Draw the graduation with a protractor and install your plumb in the right angle. That's all. In an hour or less you will have made your own emergency quadrant that can be used at any time of the night. Columbus did the same. The larger the radius of the quadrant, the more accurate your measurement will be. The famous Danish astronomer Tycho Brahe (1546-1601), renowned for his very accurate measurements, used quadrants with radii of several meters.

An example of a Polaris sight, corrected for eccentricity using the pole height correction disc

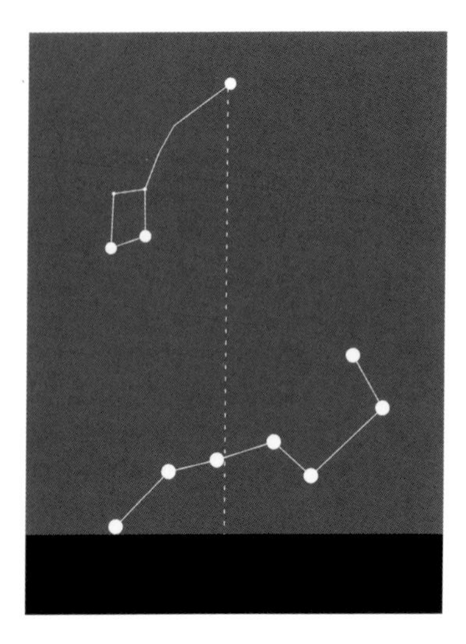

This was the northern sky seen from Gladys on the 22nd of August 1984 at 04:43 GMT. Polaris is nearly right above Alioth, the third star in the handle of the Big Dipper.

Polaris.
From measured altitude to true altitude

Measured altitude	*39° 47´.0*
Refraction	*-1´.2*
Dip horizon	*-3´.0*
Polaris, true altitude	*39° 42´.8*

At sea on Wednesday the 22nd of August 1984 I couldn't use the pole height correction disc, because it did not exist yet. Instead I worked with the "modern" Aries technique in which I calculated the local hour angle (*LHA*-Aries) between the longitude of *Gladys* and point Aries. This *LHA*-Aries I used as an entry in the "*Polaris* (*Pole Star*) Table" of the *Reeds* or the *Nautical Almanac* which gave me the true eccentricity correction for Polaris. In the box below is a detailed example of a latitude determination from a Pole Star observation, while at the Atlantic Ocean.

From a true Polaris altitude to latitude

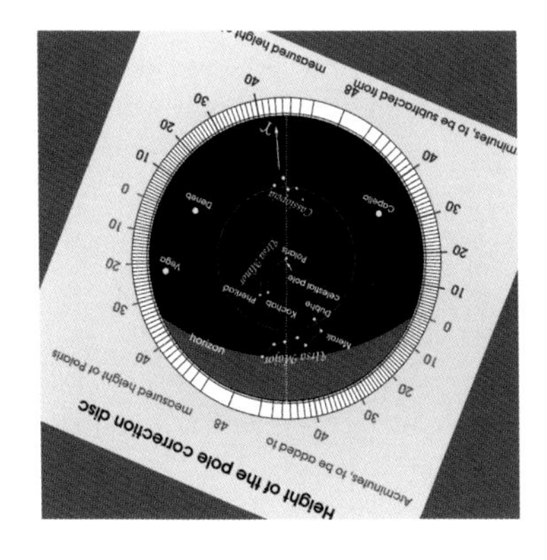

Pole height correction disc for around 1984. The disc is turned in a position that mimics the stars as seen in reality, with Polaris almost right above Alioth. Follow the vertical dotted line to 12 o'clock and you get the number 44´.5. This number (blue) should be subtracted from the true altitude of Polaris to get your latitude. Red means add, blue means subtract.

LAT = 39° 42´.8 – 44´.5 = 38° 58´.3

Calculating the altitude of the north pole using *LHA*-Aries.

Wednesday the 22nd of August 1984
Dead reckoning position: 39°N, 30°W, near the Azores;
GMT: 04 hour 43 minutes; sextant altitude Polaris (incl ic): 39°47´
QUESTION: What is the altitude of the north pole?
SOLUTION, (following the Supplements of the Reeds Almanac):

a)	*Aries GHA*	*04 Hour*	= 30° 38´.8	
	" "	*43 Minutes*	= 10° 46´.8	
			-------	*plus*
	Aries GHA 04:43		= 41° 25´.6	
	Dead reckoning longitude		30° 00´.0 W	
			-------	*minus*
	Aries Local Hour Angle		11° 25´.6	
b)	*Sextant-measured + index corr*		= 39° 47´.0	
	Corr. refraction and dip (height of eye: 3 m)		= 04´.2	*minus*

	Polaris true altitude		= 39° 42´.8	
c)	*Eccentricity corr. for LHA-Aries 11° 25´.6*		44´.5	*minus*

	True altitude of the north pole (= ships latitude)		38° 58´.3	

Explanatory note: From the *Reeds Almanac* for 04:43 GMT: hour angle difference Aries with the Greenwich meridian (*GHA*-Aries) = 41° 25´.6. That is by how much Aries was to the west of Greenwich at 04:43 GMT. According to my dead reckoning position, my longitude was 30° west. The difference is the local hour angle (*LHA*) of Aries west of my position. Thus *LHA*-Aries = 11° 25´.6. From the table "*Polaris Table for determining the latitude from a sextant altitude*" in the *Reeds Almanac 1984* I found that with this *LHA*-Aries value the eccentricity correction was -44´.5. And that feels good because it is exactly the same as also found with the 1984 pole height correction disc (see p.43), So, on Wednesday the 22 of August 1984 at 04:43 GMT the true height of the pole and hence the ship's latitude was: 38° 58´.3 N.

That same day around 18:00 GMT (20.00 h local time, in the dark) we arrived safely in Horta on the island Fayal (38° 30´ N, 28°42´ W) in the Azores.
Of course we also calculated our latitude from the meridian height of the sun on every day during this part of the journey. More about this most important navigation technique later on, during the crossing from the Azores to England.

Approaching Fayal on the Azores. Photo: DH

From the Azores to England

After five days of relaxing and preparing the boat for the next leg to England we left Horta on the 28th of August 1984. A bit late in the season, therefore: hurry up. Our next destination was Plymouth in southwest England. This was the port from where we had started for the OSTAR three months before. The distance to sail was about 1300 miles. The first days we sailed north to get into the more constant and stronger westerly winds. Then we changed course to the northeast and later to the east-northeast. See the chart on p. 37. The North Star had already climbed noticeably. And on every sunny day we took a meridian altitude of the sun to find our latitude.

Meridian altitudes

The use of meridian heights of the sun to find one's latitude anywhere at sea goes back to the 15th century. The Portuguese explorers, sailing south along the African coast in their search for the Silk Route over sea to Asia, arrived in the southern hemisphere around 1473. They could no longer use the height of Polaris, because it had disappeared below the horizon. This most annoying "latitude problem" was solved by taking solar meridian altitudes and by computing the sun's declination from the tables of Zacut or Regiomontanus. For more information, see also the chapter on the history of celestial navigation, pp. 9-13.

To calculate his latitude from a meridian altitude the navigator needs to know two variables:

1) He must know the highest true altitude of the sun of the day. At that moment the sun passes through the ship's meridian, the so called "local solar transit". It is then mid-day or noon. The bearing of the sun during local transit is directly south (or north).
2) He needs to know the sun's declination. The declination is the latitude of the point on earth where the sun is exactly straight above. Its so-called "foot point". On 21 June, for example, the declination of the sun is 23°.5 N (summer solstice). On 21 March and 23 September it is 0° because the sun is then right above the equator (spring and autumn equinox). And on 22 December the declination is 23°.5 S (winter solstice).

Already in 1492, during his westward search for the Indies, Columbus had handwritten declination tables. He was able to calculate his ship's latitude in daytime from the meridian altitude of the sun. At night he found "the height of the pole," as it was then called, from the height of the North Star, using a quadrant or an astrolabe.

To spare yourself the effort of following the rising of the sun for hours, it is useful to know at what time you can expect the "local transit" of the sun. You find out by converting your dead reckoning longitude to time. It works as follows: the Earth makes a full eastward rotation of 360° in 24 hours. So in one hour (= 60 minutes) the sun moves 15° to the west. Therefore 1° in longitude equals four minutes in time. Should your position be at 30° west of the Greenwich 0° meridian, then the sun will reach your meridian 30 x 4 = 120 minutes later than it has passed over Greenwich. The exact transit time of the sun through the 0° meridian may be found from the *Nautical Almanac*, but can also simply be looked up from the graph on p. 34. At 30° west the sun passes at around 14 hours GMT, earlier or later by at most a quarter of an hour.

The quadrant as a sundial with a perpendicular shadow pen. The plumb line reading (at the index finger) is 0° for horizontal. The measured height of the sun is here 8°. Photo: DH

Start following the still rising sun on your sextant some twenty minutes before the calculated local transit time. The rising will slow down when it approaches your meridian. When the sun rises no more, but does not yet descend either, it is in local transit, passing westward through your meridian. This is mid-day or noon. The local solar time is 12 o'clock. A few minutes later the sun begins to descend again. The afternoon has started.
The largest measured sun altitude you note down and, after having safely stored the sextant, you improve it to a true *ALT*-sun. See the example in the box on the next page. From the Almanac, and also from the tables at the back of this booklet, the declination of the sun (*DEC*) may be found. The GMT time of the declination is not so critical. An estimate of half an hour is sufficiently accurate.
Now you can start with the math, which is simple. If you are north of the sun, the solar transit is to your south and the meridian altitude formula is: $LAT = DEC + (90° - ALT)$. This gives you your ship's latitude at local noon. See also the chapter "Meridian altitude and zenith angle" (pp. 17-18).
Note that these meridian altitudes for calculating latitude at sea can be used all year round and everywhere on the world's oceans north and south of the equator. Also you don't need to know the precise time.

An example: On Wednesday the 5th of September, en route from the Azores to England, we took a meridian altitude of the sun. The details of the calculation are shown in the box below. The declination at local transit on that day was found from the *Reeds Almanac*: *DEC* = 06° 38′. The measured height of the lower limb of the Sun was, including the index correction for the sextant, *H*sext = 47° 42′. The total correction for semi-diameter, refraction and horizon dip (eye height 2.4 m) = +16′ – 1′ – 2′.7 = +12′.3. Normally, from the low position of a yacht, this total correction is around 12′-13′ (see pp. 22-24).

Therefore I mostly use just +12´.
This acceptable shortcut is named the "fisherman's correction".

Meridian altitude on Wednesday the 5 of September 1984

Sun's lower limb measured sextant altitude + ic: Hsext	*= 47° 42´*
Fisherman's correction	*+12´*
	------- +
True altitude of the Sun	*Hw = 47° 54´*
Zenith angle = 90°- ALT = 90° - 47° 54´	*= 42° 06´*
Sun's declination (Reeds Almanac)	*DEC = 06° 38´*
	------- +
Sun's meridian altitude = Ship's Latitude	*48° 44´*

Knowing your latitude is very helpful but it doesn't give you the ship's actual position. For that you also need to know the ship's longitude. For centuries finding longitude was an unsolvable problem because, although the principle was well understood, the all-important accuracy of the timekeepers at sea was insufficient. A pendulum, though very accurate when stationary on land, does not work on a pitching and rolling ship. In those days ship's longitude was estimated with the acronym LLLL.

-L1: Latitude by Polaris- and meridian altitudes. Latitude is accurate.
-L2: Log. Dead Reckoning position from log and compass. D.R. is an approximation.
-L3: Lead. Water depth under the keel. Accurate and most important when approaching land.
-L4: Look out. A sharp eye, preferably as high as practical and possible: accurate if visibility is good.
And: common-sense seamanship based on experience and local knowledge.

Our return trip to Plymouth went fine. Once in the westerly winds, we made a good daily progress of around 130 miles. Every day we measured the meridian altitude at local noon and when conditions were right we also took a Polaris altitude in the early morning. At more northerly latitudes, however, the horizon is not always sharp enough early in the morning. So we knew our daily latitude. Longitude was estimated from the log and the compass. Luckily, visibility continued to be fine as we approached England and on the 6th of September at 23 hour GMT the white flashing light (Fl W 3 sec) of Lizard Head Lighthouse (49° 57´.5 N, 005° 11´.2 W) was spotted over de port bow. A great moment again.

On the 7th of September, after ten days at sea we arrived safely in Plymouth harbor. Tired of course but very satisfied. Despite an instructive navigation crisis in the beginning of the journey, *Gladys* had twice crossed the Atlantic successfully, self-reliant and by classical navigation.
Polaris and the sun guided Columbus back to Lisbon (1493). The same Polaris and the same sun guided us back to Plymouth (1984).

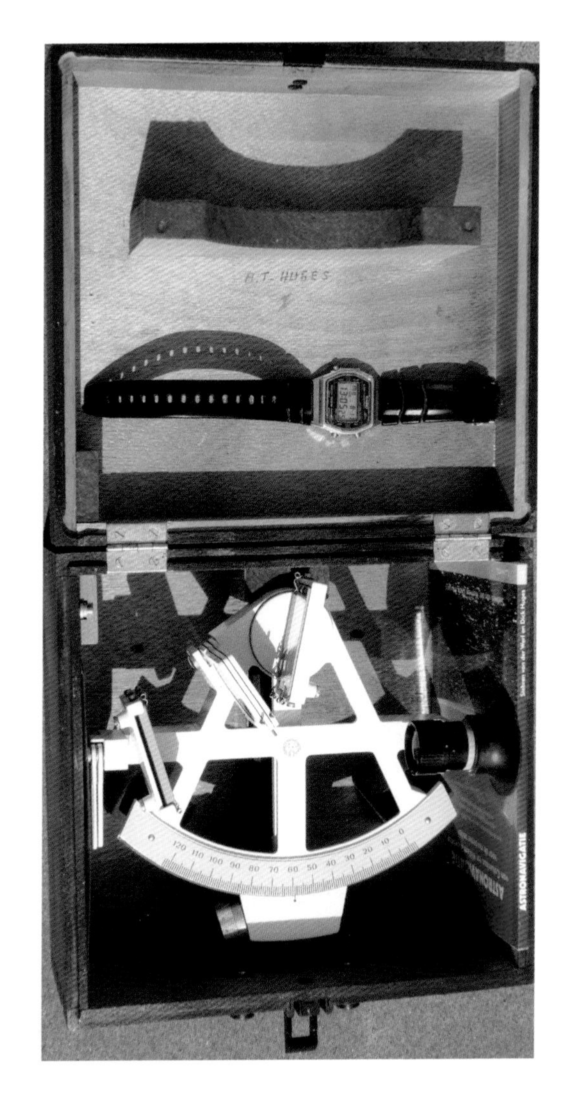

Yacht sextant with GMT-watch and this little "how-to" booklet, ready at hand in the sextant box. *Photo: DH*

Contemplations

Practical

Of course we all navigate by GPS nowadays. That is just good seamanship. But it is also good seamanship to have a backup because this fantastic e-technique can fail for several reasons. Repairs at sea will not be possible for most of us. There is a need for a "navigation life raft". A second GPS is fine of course, but it is "more of the same". An alternative is to rely on a technique, independent of electric power and e-technique, such as celestial navigation. The trip with the *Gladys* across the Atlantic in 1984 taught me that the navigation techniques of five hundred years ago still work today. All you need is a sextant (or, if necessary, you construct an emergency quadrant), a quartz clock set to GMT and this booklet that contains all the "how-to-information" you need. Including the Regiment of the North Star, the declination tables (usable for the next 20 years) and the always valid day length tables for all latitudes that give you the morning or evening longitude of your ship. In case you need emergency astronavigation, this allows you to determine a reliable position in latitude and longitude every day, wherever you are on the oceans.

Mentally

The astronavigator works with the ever-present majestic silent logic high above the boat. Doing this he joins the ranks of the classical navigators who for centuries set their course by the sun and the stars. Instead of remaining an outsider, for his position determination dependant on e-techniques of which he does not really understand the mechanism, he becomes an insider who can find his own way across the oceans with the help of his friends the Sun and the North Star. This leads to connection, peace of mind and intense satisfaction.
Astronavigation fascinates, enriches and integrates.

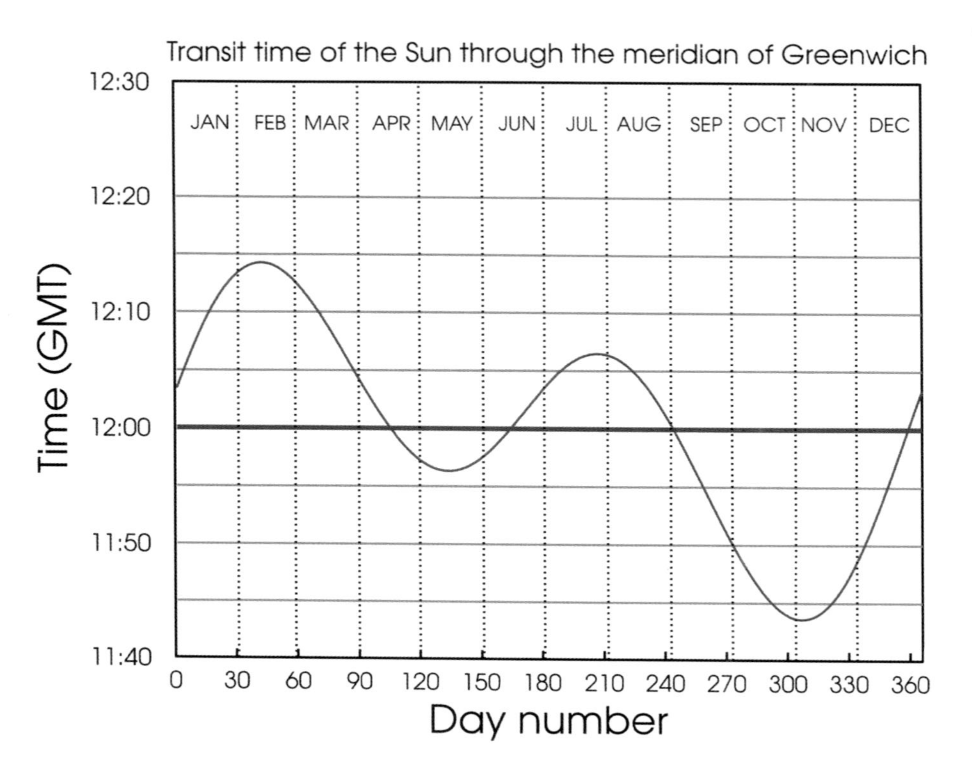
Transit time of the Sun through the meridian of Greenwich
Time (GMT)
12:30
12:20
12:10
12:00
11:50
11:40
JAN
FEB
MAR
APR
MAY
JUN
JUL
AUG
SEP
OCT
NOV
DEC
0
30
60
90
120
150
180
210
240
270
300
330
360
Day number

Day length tables for all latitudes

The tables give the daylight periods from sunrise to sunset in two versions. The first is for the sun's upper limb on the horizon, from the first light at sunrise till the last at sunset. The second version gives the times for the lower limb on the horizon, at sunrise the moment when the sun just comes free of the horizon and at sunset when its lower limb just touches the horizon again.
For the lower limb on the horizon, the true altitude of the sun is taken -20′.7 and for the upper limb -52′.7. These values correspond to a standard atmosphere and an eye height of 2.5 meters.
The tables give the day lengths for latitudes (*LAT*) of 0° - 89° in the northern hemisphere and for declinations (*DEC*) from -24° to +24°. In the southern hemisphere the corresponding daylight periods are found by reading the latitude (*LAT*) as South and looking up for a declination of the opposite sign.
An example: during the summer solstice the declination is 23° 26′. At 52°.5 N, that is for example IJmuiden on the Dutch North Sea coast and Lowestoft in England, the daylight period is then 16:51 hours. The same daylight period is seen during the winter solstice (declination = -23° 26′) at 52°.5 S, which is at the entrance to the Strait of Magellan.

UPPER LIMB

LAT = 0° - 9°

Day length (h:m) from sunrise to sunset, upper limb on horizon.

DEC (°)	LAT = 0°	1°	2°	3°	4°	5°	6°	7°	8°	9°
-24	12:08	12:04	12:01	11:57	11:53	11:50	11:46	11:43	11:39	11:35
-23	12:08	12:04	12:01	11:57	11:54	11:51	11:47	11:44	11:40	11:37
-22	12:08	12:04	12:01	11:58	11:55	11:51	11:48	11:45	11:42	11:38
-21	12:08	12:04	12:01	11:58	11:55	11:52	11:49	11:46	11:43	11:40
-20	12:07	12:05	12:02	11:59	11:56	11:53	11:50	11:47	11:44	11:41
-19	12:07	12:05	12:02	11:59	11:56	11:54	11:51	11:48	11:45	11:43
-18	12:07	12:05	12:02	12:00	11:57	11:54	11:52	11:49	11:47	11:44
-17	12:07	12:05	12:02	12:00	11:58	11:55	11:53	11:50	11:48	11:45
-16	12:07	12:05	12:03	12:00	11:58	11:56	11:54	11:51	11:49	11:47
-15	12:07	12:05	12:03	12:01	11:59	11:57	11:54	11:52	11:50	11:48
-14	12:07	12:05	12:03	12:01	11:59	11:57	11:55	11:53	11:51	11:49
-13	12:07	12:05	12:04	12:02	12:00	11:58	11:56	11:54	11:52	11:51
-12	12:07	12:05	12:04	12:02	12:00	11:59	11:57	11:55	11:54	11:52
-11	12:07	12:06	12:04	12:02	12:01	11:59	11:58	11:56	11:55	11:53
-10	12:07	12:06	12:04	12:03	12:02	12:00	11:59	11:57	11:56	11:54
-9	12:07	12:06	12:05	12:03	12:02	12:01	12:00	11:58	11:57	11:56
-8	12:07	12:06	12:05	12:04	12:03	12:01	12:00	11:59	11:58	11:57
-7	12:07	12:06	12:05	12:04	12:03	12:02	12:01	12:00	11:59	11:58
-6	12:07	12:06	12:05	12:05	12:04	12:03	12:02	12:01	12:00	12:00
-5	12:07	12:06	12:06	12:05	12:04	12:04	12:03	12:02	12:01	12:01
-4	12:07	12:06	12:06	12:05	12:05	12:04	12:04	12:03	12:03	12:02
-3	12:07	12:07	12:06	12:06	12:05	12:05	12:05	12:04	12:04	12:03
-2	12:07	12:07	12:06	12:06	12:06	12:06	12:05	12:05	12:05	12:05
-1	12:07	12:07	12:07	12:07	12:06	12:06	12:06	12:06	12:06	12:06
0	12:07	12:07	12:07	12:07	12:07	12:07	12:07	12:07	12:07	12:07
1	12:07	12:07	12:07	12:07	12:08	12:08	12:08	12:08	12:08	12:08
2	12:07	12:07	12:08	12:08	12:08	12:08	12:09	12:09	12:09	12:10
3	12:07	12:07	12:08	12:08	12:09	12:09	12:10	12:10	12:10	12:11
4	12:07	12:08	12:08	12:09	12:09	12:10	12:10	12:11	12:12	12:12
5	12:07	12:08	12:08	12:09	12:10	12:11	12:11	12:12	12:13	12:13
6	12:07	12:08	12:09	12:10	12:10	12:11	12:12	12:13	12:14	12:15
7	12:07	12:08	12:09	12:10	12:11	12:12	12:13	12:14	12:15	12:16
8	12:07	12:08	12:09	12:10	12:12	12:13	12:14	12:15	12:16	12:17
9	12:07	12:08	12:10	12:11	12:12	12:13	12:15	12:16	12:17	12:19
10	12:07	12:09	12:10	12:11	12:13	12:14	12:16	12:17	12:19	12:20
11	12:07	12:09	12:10	12:12	12:13	12:15	12:17	12:18	12:20	12:21
12	12:07	12:09	12:11	12:12	12:14	12:16	12:17	12:19	12:21	12:23
13	12:07	12:09	12:11	12:13	12:15	12:17	12:18	12:20	12:22	12:24
14	12:07	12:09	12:11	12:13	12:15	12:17	12:19	12:21	12:23	12:25
15	12:07	12:09	12:12	12:14	12:16	12:18	12:20	12:22	12:25	12:27
16	12:07	12:10	12:12	12:14	12:17	12:19	12:21	12:24	12:26	12:28
17	12:07	12:10	12:12	12:15	12:17	12:20	12:22	12:25	12:27	12:30
18	12:07	12:10	12:13	12:15	12:18	12:20	12:23	12:26	12:28	12:31
19	12:07	12:10	12:13	12:16	12:18	12:21	12:24	12:27	12:30	12:33
20	12:07	12:10	12:13	12:16	12:19	12:22	12:25	12:28	12:31	12:34
21	12:08	12:11	12:14	12:17	12:20	12:23	12:26	12:29	12:32	12:36
22	12:08	12:11	12:14	12:17	12:21	12:24	12:27	12:30	12:34	12:37
23	12:08	12:11	12:14	12:18	12:21	12:25	12:28	12:32	12:35	12:39
24	12:08	12:11	12:15	12:18	12:22	12:26	12:29	12:33	12:36	12:40

LOWER LIMB

LAT = 0° - 9°

Day length (h:m) from sunrise to sunset, lower limb on horizon.

DEC (°)	LAT = 0°	1°	2°	3°	4°	5°	6°	7°	8°	9°
-24	12:03	11:59	11:56	11:52	11:49	11:45	11:42	11:38	11:34	11:31
-23	12:03	12:00	11:56	11:53	11:49	11:46	11:43	11:39	11:36	11:32
-22	12:03	12:00	11:57	11:53	11:50	11:47	11:44	11:40	11:37	11:34
-21	12:03	12:00	11:57	11:54	11:51	11:48	11:44	11:41	11:38	11:35
-20	12:03	12:00	11:57	11:54	11:51	11:48	11:45	11:42	11:40	11:37
-19	12:03	12:00	11:57	11:55	11:52	11:49	11:46	11:44	11:41	11:38
-18	12:03	12:00	11:58	11:55	11:52	11:50	11:47	11:45	11:42	11:39
-17	12:03	12:00	11:58	11:56	11:53	11:51	11:48	11:46	11:43	11:41
-16	12:03	12:01	11:58	11:56	11:54	11:51	11:49	11:47	11:44	11:42
-15	12:03	12:01	11:59	11:56	11:54	11:52	11:50	11:48	11:46	11:43
-14	12:03	12:01	11:59	11:57	11:55	11:53	11:51	11:49	11:47	11:45
-13	12:03	12:01	11:59	11:57	11:55	11:54	11:52	11:50	11:48	11:46
-12	12:03	12:01	11:59	11:58	11:56	11:54	11:53	11:51	11:49	11:47
-11	12:03	12:01	12:00	11:58	11:57	11:55	11:53	11:52	11:50	11:49
-10	12:03	12:01	12:00	11:59	11:57	11:56	11:54	11:53	11:51	11:50
-9	12:03	12:02	12:00	11:59	11:58	11:56	11:55	11:54	11:53	11:51
-8	12:03	12:02	12:01	11:59	11:58	11:57	11:56	11:55	11:54	11:53
-7	12:03	12:02	12:01	12:00	11:59	11:58	11:57	11:56	11:55	11:54
-6	12:03	12:02	12:01	12:00	11:59	11:59	11:58	11:57	11:56	11:55
-5	12:03	12:02	12:01	12:01	12:00	11:59	11:59	11:58	11:57	11:56
-4	12:03	12:02	12:02	12:01	12:01	12:00	11:59	11:59	11:58	11:58
-3	12:03	12:02	12:02	12:02	12:01	12:01	12:00	12:00	11:59	11:59
-2	12:03	12:02	12:02	12:02	12:02	12:01	12:01	12:01	12:01	12:00
-1	12:03	12:03	12:02	12:02	12:02	12:02	12:02	12:02	12:02	12:02
0	12:03	12:03	12:03	12:03	12:03	12:03	12:03	12:03	12:03	12:03
1	12:03	12:03	12:03	12:03	12:03	12:03	12:04	12:04	12:04	12:04
2	12:03	12:03	12:03	12:04	12:04	12:04	12:04	12:05	12:05	12:05
3	12:03	12:03	12:04	12:04	12:04	12:05	12:05	12:06	12:06	12:07
4	12:03	12:03	12:04	12:04	12:05	12:06	12:06	12:07	12:07	12:08
5	12:03	12:03	12:04	12:05	12:06	12:06	12:07	12:08	12:08	12:09
6	12:03	12:04	12:04	12:05	12:06	12:07	12:08	12:09	12:10	12:10
7	12:03	12:04	12:05	12:06	12:07	12:08	12:09	12:10	12:11	12:12
8	12:03	12:04	12:05	12:06	12:07	12:08	12:10	12:11	12:12	12:13
9	12:03	12:04	12:05	12:07	12:08	12:09	12:10	12:12	12:13	12:14
10	12:03	12:04	12:06	12:07	12:08	12:10	12:11	12:13	12:14	12:16
11	12:03	12:04	12:06	12:07	12:09	12:11	12:12	12:14	12:15	12:17
12	12:03	12:05	12:06	12:08	12:10	12:11	12:13	12:15	12:17	12:18
13	12:03	12:05	12:07	12:08	12:10	12:12	12:14	12:16	12:18	12:20
14	12:03	12:05	12:07	12:09	12:11	12:13	12:15	12:17	12:19	12:21
15	12:03	12:05	12:07	12:09	12:11	12:14	12:16	12:18	12:20	12:22
16	12:03	12:05	12:07	12:10	12:12	12:14	12:17	12:19	12:21	12:24
17	12:03	12:05	12:08	12:10	12:13	12:15	12:18	12:20	12:23	12:25
18	12:03	12:06	12:08	12:11	12:13	12:16	12:19	12:21	12:24	12:27
19	12:03	12:06	12:08	12:11	12:14	12:17	12:20	12:22	12:25	12:28
20	12:03	12:06	12:09	12:12	12:15	12:18	12:20	12:23	12:26	12:29
21	12:03	12:06	12:09	12:12	12:15	12:18	12:21	12:25	12:28	12:31
22	12:03	12:06	12:09	12:13	12:16	12:19	12:22	12:26	12:29	12:32
23	12:03	12:06	12:10	12:13	12:17	12:20	12:23	12:27	12:30	12:34
24	12:03	12:07	12:10	12:14	12:17	12:21	12:24	12:28	12:32	12:35

UPPER LIMB LAT = 10° - 19°

Day length (h:m) from sunrise to sunset, upper limb on horizon.

DEC (°)	LAT = 10°	11°	12°	13°	14°	15°	16°	17°	18°	19°
-24	11:32	11:28	11:24	11:21	11:17	11:13	11:09	11:06	11:02	10:58
-23	11:33	11:30	11:26	11:23	11:19	11:16	11:12	11:08	11:05	11:01
-22	11:35	11:32	11:28	11:25	11:22	11:18	11:15	11:11	11:08	11:04
-21	11:37	11:33	11:30	11:27	11:24	11:21	11:17	11:14	11:11	11:07
-20	11:38	11:35	11:32	11:29	11:26	11:23	11:20	11:17	11:14	11:10
-19	11:40	11:37	11:34	11:31	11:28	11:25	11:22	11:19	11:16	11:13
-18	11:41	11:39	11:36	11:33	11:30	11:28	11:25	11:22	11:19	11:16
-17	11:43	11:40	11:38	11:35	11:33	11:30	11:27	11:25	11:22	11:19
-16	11:44	11:42	11:40	11:37	11:35	11:32	11:30	11:27	11:25	11:22
-15	11:46	11:44	11:41	11:39	11:37	11:35	11:32	11:30	11:28	11:25
-14	11:47	11:45	11:43	11:41	11:39	11:37	11:35	11:33	11:30	11:28
-13	11:49	11:47	11:45	11:43	11:41	11:39	11:37	11:35	11:33	11:31
-12	11:50	11:48	11:47	11:45	11:43	11:41	11:40	11:38	11:36	11:34
-11	11:52	11:50	11:48	11:47	11:45	11:44	11:42	11:40	11:39	11:37
-10	11:53	11:52	11:50	11:49	11:47	11:46	11:44	11:43	11:41	11:40
-9	11:54	11:53	11:52	11:51	11:49	11:48	11:47	11:45	11:44	11:43
-8	11:56	11:55	11:54	11:52	11:51	11:50	11:49	11:48	11:47	11:45
-7	11:57	11:56	11:55	11:54	11:53	11:52	11:51	11:50	11:49	11:48
-6	11:59	11:58	11:57	11:56	11:55	11:54	11:54	11:53	11:52	11:51
-5	12:00	11:59	11:59	11:58	11:57	11:57	11:56	11:55	11:54	11:54
-4	12:02	12:01	12:00	12:00	11:59	11:59	11:58	11:58	11:57	11:56
-3	12:03	12:02	12:02	12:02	12:01	12:01	12:00	12:00	12:00	11:59
-2	12:04	12:04	12:04	12:04	12:03	12:03	12:03	12:02	12:02	12:02
-1	12:06	12:06	12:05	12:05	12:05	12:05	12:05	12:05	12:05	12:05
0	12:07	12:07	12:07	12:07	12:07	12:07	12:07	12:07	12:07	12:07
1	12:09	12:09	12:09	12:09	12:09	12:09	12:10	12:10	12:10	12:10
2	12:10	12:10	12:11	12:11	12:11	12:12	12:12	12:12	12:13	12:13
3	12:11	12:12	12:12	12:13	12:13	12:14	12:14	12:15	12:15	12:16
4	12:13	12:13	12:14	12:15	12:15	12:16	12:17	12:17	12:18	12:18
5	12:14	12:15	12:16	12:17	12:17	12:18	12:19	12:20	12:20	12:21
6	12:16	12:17	12:17	12:18	12:19	12:20	12:21	12:22	12:23	12:24
7	12:17	12:18	12:19	12:20	12:21	12:22	12:24	12:25	12:26	12:27
8	12:19	12:20	12:21	12:22	12:23	12:25	12:26	12:27	12:28	12:30
9	12:20	12:21	12:23	12:24	12:25	12:27	12:28	12:30	12:31	12:33
10	12:22	12:23	12:24	12:26	12:28	12:29	12:31	12:32	12:34	12:35
11	12:23	12:25	12:26	12:28	12:30	12:31	12:33	12:35	12:37	12:38
12	12:24	12:26	12:28	12:30	12:32	12:34	12:35	12:37	12:39	12:41
13	12:26	12:28	12:30	12:32	12:34	12:36	12:38	12:40	12:42	12:44
14	12:28	12:30	12:32	12:34	12:36	12:38	12:40	12:43	12:45	12:47
15	12:29	12:31	12:34	12:36	12:38	12:40	12:43	12:45	12:48	12:50
16	12:31	12:33	12:35	12:38	12:40	12:43	12:45	12:48	12:50	12:53
17	12:32	12:35	12:37	12:40	12:43	12:45	12:48	12:51	12:53	12:56
18	12:34	12:37	12:39	12:42	12:45	12:48	12:50	12:53	12:56	12:59
19	12:35	12:38	12:41	12:44	12:47	12:50	12:53	12:56	12:59	13:02
20	12:37	12:40	12:43	12:46	12:49	12:53	12:56	12:59	13:02	13:06
21	12:39	12:42	12:45	12:48	12:52	12:55	12:58	13:02	13:05	13:09
22	12:40	12:44	12:47	12:51	12:54	12:58	13:01	13:05	13:08	13:12
23	12:42	12:46	12:49	12:53	12:57	13:00	13:04	13:08	13:12	13:15
24	12:44	12:48	12:51	12:55	12:59	13:03	13:07	13:11	13:15	13:19

LOWER LIMB LAT = 10° - 19°

Day length (h:m) from sunrise to sunset, lower limb on horizon.

DEC (°)	LAT = 10°	11°	12°	13°	14°	15°	16°	17°	18°	19°
-24	11:27	11:23	11:20	11:16	11:12	11:08	11:04	11:01	10:57	10:53
-23	11:29	11:25	11:22	11:18	11:15	11:11	11:07	11:04	11:00	10:56
-22	11:30	11:27	11:24	11:20	11:17	11:13	11:10	11:06	11:03	10:59
-21	11:32	11:29	11:26	11:22	11:19	11:16	11:13	11:09	11:06	11:02
-20	11:34	11:31	11:28	11:24	11:21	11:18	11:15	11:12	11:09	11:06
-19	11:35	11:32	11:29	11:27	11:24	11:21	11:18	11:15	11:12	11:09
-18	11:37	11:34	11:31	11:29	11:26	11:23	11:20	11:17	11:15	11:12
-17	11:38	11:36	11:33	11:31	11:28	11:25	11:23	11:20	11:17	11:15
-16	11:40	11:37	11:35	11:33	11:30	11:28	11:25	11:23	11:20	11:18
-15	11:41	11:39	11:37	11:35	11:32	11:30	11:28	11:25	11:23	11:21
-14	11:43	11:41	11:39	11:37	11:34	11:32	11:30	11:28	11:26	11:24
-13	11:44	11:42	11:40	11:38	11:37	11:35	11:33	11:31	11:29	11:27
-12	11:46	11:44	11:42	11:40	11:39	11:37	11:35	11:33	11:31	11:29
-11	11:47	11:46	11:44	11:42	11:41	11:39	11:37	11:36	11:34	11:32
-10	11:49	11:47	11:46	11:44	11:43	11:41	11:40	11:38	11:37	11:35
-9	11:50	11:49	11:47	11:46	11:45	11:43	11:42	11:41	11:39	11:38
-8	11:51	11:50	11:49	11:48	11:47	11:46	11:44	11:43	11:42	11:41
-7	11:53	11:52	11:51	11:50	11:49	11:48	11:47	11:46	11:45	11:44
-6	11:54	11:53	11:53	11:52	11:51	11:50	11:49	11:48	11:47	11:46
-5	11:56	11:55	11:54	11:54	11:53	11:52	11:51	11:51	11:50	11:49
-4	11:57	11:57	11:56	11:55	11:55	11:54	11:54	11:53	11:52	11:52
-3	11:59	11:58	11:58	11:57	11:57	11:56	11:56	11:56	11:55	11:55
-2	12:00	12:00	11:59	11:59	11:59	11:59	11:58	11:58	11:58	11:57
-1	12:01	12:01	12:01	12:01	12:01	12:01	12:01	12:00	12:00	12:00
0	12:03	12:03	12:03	12:03	12:03	12:03	12:03	12:03	12:03	12:03
1	12:04	12:04	12:05	12:05	12:05	12:05	12:05	12:05	12:06	12:06
2	12:06	12:06	12:06	12:07	12:07	12:07	12:07	12:08	12:08	12:08
3	12:07	12:07	12:08	12:08	12:09	12:09	12:10	12:10	12:11	12:11
4	12:08	12:09	12:10	12:10	12:11	12:11	12:12	12:13	12:13	12:14
5	12:10	12:11	12:11	12:12	12:13	12:14	12:14	12:15	12:16	12:17
6	12:11	12:12	12:13	12:14	12:15	12:16	12:17	12:18	12:19	12:20
7	12:13	12:14	12:15	12:16	12:17	12:18	12:19	12:20	12:21	12:22
8	12:14	12:15	12:17	12:18	12:19	12:20	12:21	12:23	12:24	12:25
9	12:16	12:17	12:18	12:20	12:21	12:22	12:24	12:25	12:27	12:28
10	12:17	12:19	12:20	12:22	12:23	12:25	12:26	12:28	12:29	12:31
11	12:19	12:20	12:22	12:23	12:25	12:27	12:28	12:30	12:32	12:34
12	12:20	12:22	12:24	12:25	12:27	12:29	12:31	12:33	12:35	12:37
13	12:22	12:23	12:25	12:27	12:29	12:31	12:33	12:35	12:37	12:39
14	12:23	12:25	12:27	12:29	12:31	12:34	12:36	12:38	12:40	12:42
15	12:25	12:27	12:29	12:31	12:34	12:36	12:38	12:41	12:43	12:45
16	12:26	12:28	12:31	12:33	12:36	12:38	12:41	12:43	12:46	12:48
17	12:28	12:30	12:33	12:35	12:38	12:41	12:43	12:46	12:49	12:51
18	12:29	12:32	12:35	12:37	12:40	12:43	12:46	12:49	12:52	12:54
19	12:31	12:34	12:37	12:39	12:42	12:45	12:48	12:51	12:54	12:58
20	12:32	12:35	12:39	12:42	12:45	12:48	12:51	12:54	12:57	13:01
21	12:34	12:37	12:40	12:44	12:47	12:50	12:54	12:57	13:00	13:04
22	12:36	12:39	12:42	12:46	12:49	12:53	12:56	13:00	13:04	13:07
23	12:37	12:41	12:44	12:48	12:52	12:55	12:59	13:03	13:07	13:10
24	12:39	12:43	12:47	12:50	12:54	12:58	13:02	13:06	13:10	13:14

UPPER LIMB LAT = 20° - 29°

Day length (h:m) from sunrise to sunset, upper limb on horizon.

DEC (°)	LAT = 20°	21°	22°	23°	24°	25°	26°	27°	28°	29°
-24	10:54	10:50	10:46	10:41	10:37	10:33	10:28	10:24	10:19	10:15
-23	10:57	10:53	10:49	10:45	10:41	10:37	10:33	10:29	10:25	10:20
-22	11:00	10:57	10:53	10:49	10:46	10:42	10:38	10:34	10:30	10:25
-21	11:04	11:00	10:57	10:53	10:50	10:46	10:42	10:38	10:34	10:31
-20	11:07	11:04	11:00	10:57	10:54	10:50	10:47	10:43	10:39	10:36
-19	11:10	11:07	11:04	11:01	10:58	10:54	10:51	10:48	10:44	10:41
-18	11:14	11:11	11:08	11:05	11:02	10:59	10:55	10:52	10:49	10:46
-17	11:17	11:14	11:11	11:08	11:06	11:03	11:00	10:57	10:54	10:50
-16	11:20	11:17	11:15	11:12	11:09	11:07	11:04	11:01	10:58	10:55
-15	11:23	11:21	11:18	11:16	11:13	11:11	11:08	11:05	11:03	11:00
-14	11:26	11:24	11:22	11:19	11:17	11:15	11:12	11:10	11:07	11:05
-13	11:29	11:27	11:25	11:23	11:21	11:19	11:16	11:14	11:12	11:09
-12	11:32	11:30	11:28	11:26	11:24	11:22	11:20	11:18	11:16	11:14
-11	11:35	11:33	11:32	11:30	11:28	11:26	11:24	11:23	11:21	11:19
-10	11:38	11:37	11:35	11:33	11:32	11:30	11:28	11:27	11:25	11:23
-9	11:41	11:40	11:38	11:37	11:35	11:34	11:32	11:31	11:29	11:28
-8	11:44	11:43	11:42	11:40	11:39	11:38	11:36	11:35	11:34	11:32
-7	11:47	11:46	11:45	11:44	11:43	11:42	11:40	11:39	11:38	11:37
-6	11:50	11:49	11:48	11:47	11:46	11:45	11:44	11:43	11:42	11:41
-5	11:53	11:52	11:51	11:51	11:50	11:49	11:48	11:47	11:47	11:46
-4	11:56	11:55	11:55	11:54	11:53	11:53	11:52	11:52	11:51	11:50
-3	11:59	11:58	11:58	11:57	11:57	11:57	11:56	11:56	11:55	11:55
-2	12:02	12:01	12:01	12:01	12:01	12:00	12:00	12:00	11:59	11:59
-1	12:05	12:04	12:04	12:04	12:04	12:04	12:04	12:04	12:04	12:04
0	12:07	12:08	12:08	12:08	12:08	12:08	12:08	12:08	12:08	12:08
1	12:10	12:11	12:11	12:11	12:11	12:11	12:12	12:12	12:12	12:12
2	12:13	12:14	12:14	12:14	12:15	12:15	12:16	12:16	12:16	12:17
3	12:16	12:17	12:17	12:18	12:18	12:19	12:20	12:20	12:21	12:21
4	12:19	12:20	12:21	12:21	12:22	12:23	12:23	12:24	12:25	12:26
5	12:22	12:23	12:24	12:25	12:26	12:26	12:27	12:28	12:29	12:30
6	12:25	12:26	12:27	12:28	12:29	12:30	12:31	12:33	12:34	12:35
7	12:28	12:29	12:30	12:32	12:33	12:34	12:35	12:37	12:38	12:39
8	12:31	12:32	12:34	12:35	12:36	12:38	12:39	12:41	12:42	12:44
9	12:34	12:36	12:37	12:39	12:40	12:42	12:43	12:45	12:47	12:48
10	12:37	12:39	12:40	12:42	12:44	12:46	12:47	12:49	12:51	12:53
11	12:40	12:42	12:44	12:46	12:48	12:50	12:52	12:54	12:56	12:58
12	12:43	12:45	12:47	12:49	12:51	12:53	12:56	12:58	13:00	13:02
13	12:46	12:48	12:51	12:53	12:55	12:57	13:00	13:02	13:05	13:07
14	12:49	12:52	12:54	12:57	12:59	13:01	13:04	13:07	13:09	13:12
15	12:53	12:55	12:58	13:00	13:03	13:06	13:08	13:11	13:14	13:17
16	12:56	12:58	13:01	13:04	13:07	13:10	13:13	13:16	13:19	13:22
17	12:59	13:02	13:05	13:08	13:11	13:14	13:17	13:20	13:23	13:27
18	13:02	13:05	13:08	13:12	13:15	13:18	13:21	13:25	13:28	13:32
19	13:06	13:09	13:12	13:15	13:19	13:22	13:26	13:29	13:33	13:37
20	13:09	13:12	13:16	13:19	13:23	13:27	13:30	13:34	13:38	13:42
21	13:12	13:16	13:20	13:23	13:27	13:31	13:35	13:39	13:43	13:47
22	13:16	13:20	13:23	13:27	13:31	13:35	13:40	13:44	13:48	13:52
23	13:19	13:23	13:27	13:31	13:36	13:40	13:44	13:49	13:53	13:58
24	13:23	13:27	13:31	13:36	13:40	13:45	13:49	13:54	13:59	14:03

LOWER LIMB LAT = 20° - 29°

Day length (h:m) from sunrise to sunset, lower limb on horizon.

DEC (°)	LAT = 20°	21°	22°	23°	24°	25°	26°	27°	28°	29°
-24	10:49	10:45	10:40	10:36	10:32	10:28	10:23	10:19	10:14	10:09
-23	10:52	10:48	10:44	10:40	10:36	10:32	10:28	10:24	10:19	10:15
-22	10:56	10:52	10:48	10:44	10:40	10:36	10:32	10:28	10:24	10:20
-21	10:59	10:55	10:52	10:48	10:45	10:41	10:37	10:33	10:29	10:25
-20	11:02	10:59	10:56	10:52	10:49	10:45	10:42	10:38	10:34	10:30
-19	11:06	11:02	10:59	10:56	10:53	10:49	10:46	10:42	10:39	10:35
-18	11:09	11:06	11:03	11:00	10:57	10:54	10:50	10:47	10:44	10:40
-17	11:12	11:09	11:06	11:04	11:01	10:58	10:55	10:52	10:48	10:45
-16	11:15	11:13	11:10	11:07	11:04	11:02	10:59	10:56	10:53	10:50
-15	11:18	11:16	11:13	11:11	11:08	11:06	11:03	11:00	10:58	10:55
-14	11:21	11:19	11:17	11:15	11:12	11:10	11:07	11:05	11:02	11:00
-13	11:24	11:22	11:20	11:18	11:16	11:14	11:11	11:09	11:07	11:04
-12	11:28	11:26	11:24	11:22	11:20	11:18	11:16	11:13	11:11	11:09
-11	11:31	11:29	11:27	11:25	11:23	11:22	11:20	11:18	11:16	11:14
-10	11:34	11:32	11:30	11:29	11:27	11:25	11:24	11:22	11:20	11:18
-9	11:37	11:35	11:34	11:32	11:31	11:29	11:28	11:26	11:25	11:23
-8	11:40	11:38	11:37	11:36	11:34	11:33	11:32	11:30	11:29	11:27
-7	11:42	11:41	11:40	11:39	11:38	11:37	11:36	11:34	11:33	11:32
-6	11:45	11:44	11:44	11:43	11:42	11:41	11:40	11:39	11:38	11:36
-5	11:48	11:48	11:47	11:46	11:45	11:44	11:44	11:43	11:42	11:41
-4	11:51	11:51	11:50	11:49	11:49	11:48	11:47	11:47	11:46	11:45
-3	11:54	11:54	11:53	11:53	11:52	11:52	11:51	11:51	11:50	11:50
-2	11:57	11:57	11:57	11:56	11:56	11:56	11:55	11:55	11:55	11:54
-1	12:00	12:00	12:00	12:00	11:59	11:59	11:59	11:59	11:59	11:59
0	12:03	12:03	12:03	12:03	12:03	12:03	12:03	12:03	12:03	12:03
1	12:06	12:06	12:06	12:06	12:07	12:07	12:07	12:07	12:07	12:08
2	12:09	12:09	12:09	12:10	12:10	12:11	12:11	12:11	12:12	12:12
3	12:12	12:12	12:13	12:13	12:14	12:14	12:15	12:15	12:16	12:16
4	12:15	12:15	12:16	12:17	12:17	12:18	12:19	12:19	12:20	12:21
5	12:18	12:18	12:19	12:20	12:21	12:22	12:23	12:24	12:24	12:25
6	12:20	12:21	12:22	12:23	12:24	12:26	12:27	12:28	12:29	12:30
7	12:23	12:25	12:26	12:27	12:28	12:29	12:31	12:32	12:33	12:34
8	12:26	12:28	12:29	12:30	12:32	12:33	12:35	12:36	12:37	12:39
9	12:29	12:31	12:32	12:34	12:35	12:37	12:39	12:40	12:42	12:44
10	12:32	12:34	12:36	12:37	12:39	12:41	12:43	12:44	12:46	12:48
11	12:35	12:37	12:39	12:41	12:43	12:45	12:47	12:49	12:51	12:53
12	12:39	12:40	12:42	12:44	12:47	12:49	12:51	12:53	12:55	12:57
13	12:42	12:44	12:46	12:48	12:50	12:53	12:55	12:57	13:00	13:02
14	12:45	12:47	12:49	12:52	12:54	12:57	12:59	13:02	13:04	13:07
15	12:48	12:50	12:53	12:55	12:58	13:01	13:03	13:06	13:09	13:12
16	12:51	12:54	12:56	12:59	13:02	13:05	13:08	13:10	13:13	13:16
17	12:54	12:57	13:00	13:03	13:06	13:09	13:12	13:15	13:18	13:21
18	12:57	13:00	13:04	13:07	13:10	13:13	13:16	13:20	13:23	13:26
19	13:01	13:04	13:07	13:10	13:14	13:17	13:21	13:24	13:28	13:31
20	13:04	13:07	13:11	13:14	13:18	13:21	13:25	13:29	13:33	13:37
21	13:07	13:11	13:15	13:18	13:22	13:26	13:30	13:34	13:38	13:42
22	13:11	13:15	13:18	13:22	13:26	13:30	13:34	13:38	13:43	13:47
23	13:14	13:18	13:22	13:26	13:30	13:35	13:39	13:43	13:48	13:52
24	13:18	13:22	13:26	13:30	13:35	13:39	13:44	13:48	13:53	13:58

UPPER LIMB LAT = 30° - 39°

Day length (h:m) from sunrise to sunset, upper limb on horizon.

DEC (°)	LAT = 30°	31°	32°	33°	34°	35°	36°	37°	38°	39°
-24	10:10	10:05	10:00	9:55	9:50	9:45	9:39	9:33	9:28	9:21
-23	10:16	10:11	10:06	10:01	9:56	9:51	9:46	9:41	9:35	9:30
-22	10:21	10:17	10:12	10:08	10:03	9:58	9:53	9:48	9:43	9:38
-21	10:26	10:22	10:18	10:14	10:09	10:05	10:00	9:55	9:50	9:45
-20	10:32	10:28	10:24	10:20	10:16	10:11	10:07	10:02	9:58	9:53
-19	10:37	10:33	10:30	10:26	10:22	10:18	10:14	10:09	10:05	10:00
-18	10:42	10:39	10:35	10:32	10:28	10:24	10:20	10:16	10:12	10:08
-17	10:47	10:44	10:41	10:37	10:34	10:30	10:27	10:23	10:19	10:15
-16	10:52	10:49	10:46	10:43	10:40	10:36	10:33	10:30	10:26	10:22
-15	10:57	10:54	10:52	10:49	10:46	10:43	10:39	10:36	10:33	10:29
-14	11:02	11:00	10:57	10:54	10:51	10:49	10:46	10:43	10:39	10:36
-13	11:07	11:05	11:02	11:00	10:57	10:54	10:52	10:49	10:46	10:43
-12	11:12	11:10	11:07	11:05	11:03	11:00	10:58	10:55	10:53	10:50
-11	11:17	11:15	11:13	11:11	11:08	11:06	11:04	11:02	10:59	10:57
-10	11:22	11:20	11:18	11:16	11:14	11:12	11:10	11:08	11:06	11:04
-9	11:26	11:25	11:23	11:21	11:20	11:18	11:16	11:14	11:12	11:10
-8	11:31	11:30	11:28	11:27	11:25	11:24	11:22	11:20	11:19	11:17
-7	11:36	11:34	11:33	11:32	11:31	11:29	11:28	11:26	11:25	11:23
-6	11:40	11:39	11:38	11:37	11:36	11:35	11:34	11:33	11:31	11:30
-5	11:45	11:44	11:43	11:42	11:41	11:41	11:40	11:39	11:38	11:37
-4	11:50	11:49	11:48	11:48	11:47	11:46	11:45	11:45	11:44	11:43
-3	11:54	11:54	11:53	11:53	11:52	11:52	11:51	11:51	11:50	11:50
-2	11:59	11:59	11:58	11:58	11:58	11:57	11:57	11:57	11:56	11:56
-1	12:03	12:03	12:03	12:03	12:03	12:03	12:03	12:03	12:03	12:03
0	12:08	12:08	12:08	12:08	12:08	12:09	12:09	12:09	12:09	12:09
1	12:13	12:13	12:13	12:14	12:14	12:14	12:15	12:15	12:15	12:16
2	12:17	12:18	12:18	12:19	12:19	12:20	12:20	12:21	12:21	12:22
3	12:22	12:23	12:23	12:24	12:25	12:25	12:26	12:27	12:28	12:29
4	12:27	12:27	12:28	12:29	12:30	12:31	12:32	12:33	12:34	12:35
5	12:31	12:32	12:33	12:34	12:36	12:37	12:38	12:39	12:40	12:42
6	12:36	12:37	12:38	12:40	12:41	12:42	12:44	12:45	12:47	12:48
7	12:41	12:42	12:44	12:45	12:47	12:48	12:50	12:51	12:53	12:55
8	12:45	12:47	12:49	12:50	12:52	12:54	12:56	12:58	13:00	13:01
9	12:50	12:52	12:54	12:56	12:58	13:00	13:02	13:04	13:06	13:08
10	12:55	12:57	12:59	13:01	13:03	13:06	13:08	13:10	13:13	13:15
11	13:00	13:02	13:04	13:07	13:09	13:11	13:14	13:16	13:19	13:22
12	13:05	13:07	13:10	13:12	13:15	13:17	13:20	13:23	13:26	13:29
13	13:10	13:12	13:15	13:18	13:20	13:23	13:26	13:29	13:32	13:36
14	13:15	13:17	13:20	13:23	13:26	13:29	13:33	13:36	13:39	13:43
15	13:20	13:23	13:26	13:29	13:32	13:36	13:39	13:43	13:46	13:50
16	13:25	13:28	13:31	13:35	13:38	13:42	13:45	13:49	13:53	13:57
17	13:30	13:33	13:37	13:41	13:44	13:48	13:52	13:56	14:00	14:04
18	13:35	13:39	13:43	13:46	13:50	13:54	13:59	14:03	14:07	14:12
19	13:41	13:44	13:48	13:52	13:57	14:01	14:05	14:10	14:15	14:20
20	13:46	13:50	13:54	13:59	14:03	14:08	14:12	14:17	14:22	14:27
21	13:51	13:56	14:00	14:05	14:09	14:14	14:19	14:24	14:30	14:35
22	13:57	14:02	14:06	14:11	14:16	14:21	14:26	14:32	14:37	14:43
23	14:03	14:07	14:12	14:18	14:23	14:28	14:34	14:39	14:45	14:51
24	14:08	14:13	14:19	14:24	14:30	14:35	14:41	14:47	14:53	15:00

LOWER LIMB LAT = 30° - 39°

Day length (h:m) from sunrise to sunset, lower limb on horizon.

DEC (°)	LAT = 30°	31°	32°	33°	34°	35°	36°	37°	38°	39°
-24	10:04	10:00	9:54	9:49	9:44	9:39	9:33	9:27	9:21	9:15
-23	10:10	10:05	10:01	9:56	9:51	9:46	9:40	9:35	9:29	9:23
-22	10:16	10:11	10:07	10:02	9:57	9:52	9:47	9:42	9:37	9:31
-21	10:21	10:17	10:13	10:08	10:04	9:59	9:54	9:49	9:44	9:39
-20	10:26	10:22	10:18	10:14	10:10	10:06	10:01	9:56	9:52	9:47
-19	10:32	10:28	10:24	10:20	10:16	10:12	10:08	10:03	9:59	9:54
-18	10:37	10:33	10:30	10:26	10:22	10:18	10:14	10:10	10:06	10:02
-17	10:42	10:39	10:35	10:32	10:28	10:25	10:21	10:17	10:13	10:09
-16	10:47	10:44	10:41	10:38	10:34	10:31	10:27	10:24	10:20	10:16
-15	10:52	10:49	10:46	10:43	10:40	10:37	10:34	10:30	10:27	10:24
-14	10:57	10:54	10:52	10:49	10:46	10:43	10:40	10:37	10:34	10:31
-13	11:02	11:00	10:57	10:54	10:52	10:49	10:46	10:43	10:41	10:38
-12	11:07	11:05	11:02	11:00	10:57	10:55	10:52	10:50	10:47	10:44
-11	11:12	11:10	11:08	11:05	11:03	11:01	10:59	10:56	10:54	10:51
-10	11:17	11:15	11:13	11:11	11:09	11:07	11:05	11:02	11:00	10:58
-9	11:21	11:20	11:18	11:16	11:14	11:12	11:11	11:09	11:07	11:05
-8	11:26	11:25	11:23	11:21	11:20	11:18	11:17	11:15	11:13	11:11
-7	11:31	11:29	11:28	11:27	11:25	11:24	11:23	11:21	11:20	11:18
-6	11:35	11:34	11:33	11:32	11:31	11:30	11:28	11:27	11:26	11:25
-5	11:40	11:39	11:38	11:37	11:36	11:35	11:34	11:33	11:32	11:31
-4	11:45	11:44	11:43	11:42	11:42	11:41	11:40	11:39	11:38	11:38
-3	11:49	11:49	11:48	11:48	11:47	11:47	11:46	11:45	11:45	11:44
-2	11:54	11:54	11:53	11:53	11:53	11:52	11:52	11:51	11:51	11:51
-1	11:59	11:58	11:58	11:58	11:58	11:58	11:58	11:57	11:57	11:57
0	12:03	12:03	12:03	12:03	12:03	12:03	12:03	12:03	12:04	12:04
1	12:08	12:08	12:08	12:08	12:09	12:09	12:09	12:09	12:10	12:10
2	12:12	12:13	12:13	12:14	12:14	12:15	12:15	12:16	12:16	12:17
3	12:17	12:18	12:18	12:19	12:20	12:20	12:21	12:22	12:22	12:23
4	12:22	12:22	12:23	12:24	12:25	12:26	12:27	12:28	12:29	12:30
5	12:26	12:27	12:28	12:29	12:30	12:31	12:33	12:34	12:35	12:36
6	12:31	12:32	12:33	12:35	12:36	12:37	12:38	12:40	12:41	12:43
7	12:36	12:37	12:38	12:40	12:41	12:43	12:44	12:46	12:48	12:49
8	12:40	12:42	12:44	12:45	12:47	12:49	12:50	12:52	12:54	12:56
9	12:45	12:47	12:49	12:51	12:52	12:54	12:56	12:58	13:00	13:03
10	12:50	12:52	12:54	12:56	12:58	13:00	13:02	13:05	13:07	13:09
11	12:55	12:57	12:59	13:01	13:04	13:06	13:08	13:11	13:13	13:16
12	13:00	13:02	13:04	13:07	13:09	13:12	13:15	13:17	13:20	13:23
13	13:05	13:07	13:10	13:12	13:15	13:18	13:21	13:24	13:27	13:30
14	13:10	13:12	13:15	13:18	13:21	13:24	13:27	13:30	13:34	13:37
15	13:15	13:17	13:21	13:24	13:27	13:30	13:33	13:37	13:40	13:44
16	13:20	13:23	13:26	13:29	13:33	13:36	13:40	13:44	13:47	13:51
17	13:25	13:28	13:32	13:35	13:39	13:43	13:46	13:50	13:54	13:59
18	13:30	13:34	13:37	13:41	13:45	13:49	13:53	13:57	14:01	14:06
19	13:35	13:39	13:43	13:47	13:51	13:55	14:00	14:04	14:09	14:13
20	13:41	13:45	13:49	13:53	13:57	14:02	14:06	14:11	14:16	14:21
21	13:46	13:50	13:55	13:59	14:04	14:08	14:13	14:18	14:24	14:29
22	13:51	13:56	14:01	14:05	14:10	14:15	14:20	14:26	14:31	14:37
23	13:57	14:02	14:07	14:12	14:17	14:22	14:28	14:33	14:39	14:45
24	14:03	14:08	14:13	14:18	14:24	14:29	14:35	14:41	14:47	14:53

UPPER LIMB LAT = 40° - 49°

Day length (h:m) from sunrise to sunset, upper limb on horizon.

DEC (°)	LAT = 40°	41°	42°	43°	44°	45°	46°	47°	48°	49°
-24	9:15	9:09	9:02	8:55	8:48	8:41	8:33	8:25	8:16	8:07
-23	9:24	9:18	9:11	9:05	8:58	8:51	8:44	8:36	8:28	8:19
-22	9:32	9:26	9:20	9:14	9:08	9:01	8:54	8:47	8:39	8:31
-21	9:40	9:35	9:29	9:23	9:17	9:11	9:04	8:58	8:50	8:43
-20	9:48	9:43	9:38	9:32	9:26	9:21	9:14	9:08	9:01	8:54
-19	9:56	9:51	9:46	9:41	9:36	9:30	9:24	9:18	9:12	9:06
-18	10:03	9:59	9:54	9:49	9:44	9:39	9:34	9:28	9:23	9:16
-17	10:11	10:07	10:02	9:58	9:53	9:48	9:43	9:38	9:33	9:27
-16	10:18	10:15	10:10	10:06	10:02	9:57	9:53	9:48	9:43	9:38
-15	10:26	10:22	10:18	10:14	10:10	10:06	10:02	9:58	9:53	9:48
-14	10:33	10:30	10:26	10:23	10:19	10:15	10:11	10:07	10:03	9:58
-13	10:40	10:37	10:34	10:31	10:27	10:24	10:20	10:16	10:12	10:08
-12	10:47	10:44	10:42	10:39	10:35	10:32	10:29	10:25	10:22	10:18
-11	10:54	10:52	10:49	10:46	10:44	10:41	10:38	10:34	10:31	10:28
-10	11:01	10:59	10:57	10:54	10:52	10:49	10:46	10:43	10:41	10:37
-9	11:08	11:06	11:04	11:02	11:00	10:57	10:55	10:52	10:50	10:47
-8	11:15	11:13	11:11	11:10	11:08	11:05	11:03	11:01	10:59	10:57
-7	11:22	11:20	11:19	11:17	11:15	11:14	11:12	11:10	11:08	11:06
-6	11:29	11:27	11:26	11:25	11:23	11:22	11:20	11:19	11:17	11:15
-5	11:36	11:34	11:33	11:32	11:31	11:30	11:29	11:27	11:26	11:25
-4	11:42	11:41	11:41	11:40	11:39	11:38	11:37	11:36	11:35	11:34
-3	11:49	11:48	11:48	11:47	11:47	11:46	11:45	11:45	11:44	11:43
-2	11:56	11:55	11:55	11:55	11:54	11:54	11:54	11:53	11:53	11:52
-1	12:02	12:02	12:02	12:02	12:02	12:02	12:02	12:02	12:02	12:02
0	12:09	12:09	12:09	12:10	12:10	12:10	12:10	12:10	12:11	12:11
1	12:16	12:16	12:17	12:17	12:17	12:18	12:18	12:19	12:19	12:20
2	12:23	12:23	12:24	12:25	12:25	12:26	12:27	12:27	12:28	12:29
3	12:29	12:30	12:31	12:32	12:33	12:34	12:35	12:36	12:37	12:38
4	12:36	12:37	12:38	12:40	12:41	12:42	12:43	12:45	12:46	12:48
5	12:43	12:44	12:46	12:47	12:49	12:50	12:52	12:53	12:55	12:57
6	12:50	12:51	12:53	12:55	12:56	12:58	13:00	13:02	13:04	13:06
7	12:57	12:58	13:00	13:02	13:04	13:07	13:09	13:11	13:13	13:16
8	13:04	13:06	13:08	13:10	13:12	13:15	13:17	13:20	13:23	13:25
9	13:10	13:13	13:15	13:18	13:20	13:23	13:26	13:29	13:32	13:35
10	13:18	13:20	13:23	13:26	13:29	13:32	13:35	13:38	13:41	13:45
11	13:25	13:27	13:30	13:34	13:37	13:40	13:43	13:47	13:51	13:55
12	13:32	13:35	13:38	13:42	13:45	13:49	13:52	13:56	14:00	14:05
13	13:39	13:42	13:46	13:50	13:53	13:57	14:01	14:06	14:10	14:15
14	13:46	13:50	13:54	13:58	14:02	14:06	14:11	14:15	14:20	14:25
15	13:54	13:58	14:02	14:06	14:10	14:15	14:20	14:25	14:30	14:35
16	14:01	14:06	14:10	14:14	14:19	14:24	14:29	14:35	14:40	14:46
17	14:09	14:13	14:18	14:23	14:28	14:33	14:39	14:45	14:51	14:57
18	14:17	14:21	14:27	14:32	14:37	14:43	14:49	14:55	15:01	15:08
19	14:25	14:30	14:35	14:41	14:46	14:52	14:59	15:05	15:12	15:19
20	14:33	14:38	14:44	14:50	14:56	15:02	15:09	15:16	15:23	15:31
21	14:41	14:47	14:53	14:59	15:05	15:12	15:19	15:27	15:34	15:43
22	14:49	14:55	15:02	15:08	15:15	15:22	15:30	15:38	15:46	15:55
23	14:58	15:04	15:11	15:18	15:25	15:33	15:41	15:49	15:58	16:07
24	15:06	15:13	15:20	15:28	15:36	15:44	15:52	16:01	16:10	16:20

LOWER LIMB LAT = 40° - 49°

Day length (h:m) from sunrise to sunset, lower limb on horizon.

DEC (°)	LAT = 40°	41°	42°	43°	44°	45°	46°	47°	48°	49°
-24	9:09	9:02	8:55	8:48	8:41	8:33	8:25	8:17	8:08	7:59
-23	9:17	9:11	9:05	8:58	8:51	8:44	8:36	8:28	8:20	8:11
-22	9:26	9:20	9:14	9:07	9:01	8:54	8:47	8:39	8:32	8:24
-21	9:34	9:28	9:22	9:17	9:10	9:04	8:57	8:50	8:43	8:35
-20	9:42	9:37	9:31	9:26	9:20	9:14	9:07	9:01	8:54	8:47
-19	9:50	9:45	9:40	9:34	9:29	9:23	9:17	9:11	9:05	8:58
-18	9:57	9:53	9:48	9:43	9:38	9:33	9:27	9:21	9:15	9:09
-17	10:05	10:01	9:56	9:52	9:47	9:42	9:37	9:31	9:26	9:20
-16	10:12	10:08	10:04	10:00	9:56	9:51	9:46	9:41	9:36	9:31
-15	10:20	10:16	10:12	10:08	10:04	10:00	9:55	9:51	9:46	9:41
-14	10:27	10:24	10:20	10:16	10:13	10:09	10:05	10:00	9:56	9:51
-13	10:34	10:31	10:28	10:25	10:21	10:17	10:14	10:10	10:06	10:01
-12	10:42	10:39	10:36	10:32	10:29	10:26	10:22	10:19	10:15	10:11
-11	10:49	10:46	10:43	10:40	10:37	10:34	10:31	10:28	10:25	10:21
-10	10:56	10:53	10:51	10:48	10:46	10:43	10:40	10:37	10:34	10:31
-9	11:03	11:00	10:58	10:56	10:54	10:51	10:49	10:46	10:43	10:40
-8	11:09	11:08	11:06	11:04	11:02	10:59	10:57	10:55	10:52	10:50
-7	11:16	11:15	11:13	11:11	11:09	11:08	11:06	11:04	11:01	10:59
-6	11:23	11:22	11:20	11:19	11:17	11:16	11:14	11:12	11:11	11:09
-5	11:30	11:29	11:28	11:26	11:25	11:24	11:22	11:21	11:20	11:18
-4	11:37	11:36	11:35	11:34	11:33	11:32	11:31	11:30	11:29	11:27
-3	11:43	11:43	11:42	11:41	11:41	11:40	11:39	11:38	11:37	11:37
-2	11:50	11:50	11:49	11:49	11:48	11:48	11:47	11:47	11:46	11:46
-1	11:57	11:57	11:57	11:56	11:56	11:56	11:56	11:55	11:55	11:55
0	12:04	12:04	12:04	12:04	12:04	12:04	12:04	12:04	12:04	12:04
1	12:10	12:11	12:11	12:11	12:12	12:12	12:12	12:13	12:13	12:13
2	12:17	12:18	12:18	12:19	12:19	12:20	12:21	12:21	12:22	12:23
3	12:24	12:25	12:25	12:26	12:27	12:28	12:29	12:30	12:31	12:32
4	12:31	12:32	12:33	12:34	12:35	12:36	12:37	12:38	12:40	12:41
5	12:37	12:39	12:40	12:41	12:43	12:44	12:46	12:47	12:49	12:50
6	12:44	12:46	12:47	12:49	12:50	12:52	12:54	12:56	12:58	13:00
7	12:51	12:53	12:55	12:56	12:58	13:00	13:02	13:05	13:07	13:09
8	12:58	13:00	13:02	13:04	13:06	13:09	13:11	13:13	13:16	13:19
9	13:05	13:07	13:09	13:12	13:14	13:17	13:20	13:22	13:25	13:28
10	13:12	13:14	13:17	13:20	13:22	13:25	13:28	13:31	13:35	13:38
11	13:19	13:22	13:24	13:27	13:31	13:34	13:37	13:40	13:44	13:48
12	13:26	13:29	13:32	13:35	13:39	13:42	13:46	13:50	13:54	13:58
13	13:33	13:36	13:40	13:43	13:47	13:51	13:55	13:59	14:03	14:08
14	13:40	13:44	13:48	13:52	13:56	14:00	14:04	14:08	14:13	14:18
15	13:48	13:52	13:56	14:00	14:04	14:09	14:13	14:18	14:23	14:28
16	13:55	13:59	14:04	14:08	14:13	14:18	14:23	14:28	14:33	14:39
17	14:03	14:07	14:12	14:17	14:22	14:27	14:32	14:38	14:43	14:49
18	14:11	14:15	14:20	14:25	14:31	14:36	14:42	14:48	14:54	15:00
19	14:18	14:23	14:29	14:34	14:40	14:46	14:52	14:58	15:05	15:12
20	14:26	14:32	14:37	14:43	14:49	14:55	15:02	15:08	15:16	15:23
21	14:34	14:40	14:46	14:52	14:59	15:05	15:12	15:19	15:27	15:35
22	14:43	14:49	14:55	15:01	15:08	15:15	15:23	15:30	15:38	15:47
23	14:51	14:58	15:04	15:11	15:18	15:26	15:33	15:42	15:50	15:59
24	15:00	15:07	15:14	15:21	15:28	15:36	15:45	15:53	16:02	16:12

UPPER LIMB LAT = 50° - 59°

Day length (h:m) from sunrise to sunset, upper limb on horizon.

DEC (°)	LAT = 50°	51°	52°	53°	54°	55°	56°	57°	58°	59°
-24	7:58	7:48	7:37	7:26	7:14	7:01	6:48	6:33	6:17	5:59
-23	8:11	8:01	7:51	7:41	7:30	7:18	7:05	6:52	6:37	6:21
-22	8:23	8:14	8:05	7:55	7:45	7:34	7:22	7:10	6:56	6:41
-21	8:35	8:27	8:18	8:09	8:00	7:50	7:39	7:27	7:15	7:01
-20	8:47	8:40	8:31	8:23	8:14	8:05	7:54	7:44	7:32	7:20
-19	8:59	8:52	8:44	8:36	8:28	8:19	8:10	8:00	7:49	7:38
-18	9:10	9:03	8:57	8:49	8:41	8:33	8:25	8:15	8:06	7:55
-17	9:21	9:15	9:09	9:02	8:55	8:47	8:39	8:30	8:21	8:12
-16	9:32	9:26	9:20	9:14	9:07	9:00	8:53	8:45	8:37	8:28
-15	9:43	9:38	9:32	9:26	9:20	9:14	9:07	9:00	8:52	8:44
-14	9:53	9:49	9:43	9:38	9:32	9:27	9:20	9:14	9:07	8:59
-13	10:04	9:59	9:55	9:50	9:45	9:39	9:34	9:27	9:21	9:14
-12	10:14	10:10	10:06	10:01	9:57	9:52	9:47	9:41	9:35	9:29
-11	10:24	10:21	10:17	10:13	10:08	10:04	9:59	9:54	9:49	9:44
-10	10:34	10:31	10:28	10:24	10:20	10:16	10:12	10:08	10:03	9:58
-9	10:44	10:41	10:38	10:35	10:32	10:28	10:24	10:20	10:16	10:12
-8	10:54	10:51	10:49	10:46	10:43	10:40	10:37	10:33	10:30	10:26
-7	11:04	11:02	10:59	10:57	10:54	10:52	10:49	10:46	10:43	10:40
-6	11:13	11:12	11:10	11:08	11:06	11:03	11:01	10:59	10:56	10:53
-5	11:23	11:22	11:20	11:18	11:17	11:15	11:13	11:11	11:09	11:07
-4	11:33	11:32	11:30	11:29	11:28	11:26	11:25	11:24	11:22	11:20
-3	11:42	11:42	11:41	11:40	11:39	11:38	11:37	11:36	11:35	11:34
-2	11:52	11:51	11:51	11:50	11:50	11:49	11:49	11:48	11:48	11:47
-1	12:01	12:01	12:01	12:01	12:01	12:01	12:01	12:01	12:00	12:00
0	12:11	12:11	12:11	12:12	12:12	12:12	12:13	12:13	12:13	12:14
1	12:20	12:21	12:22	12:22	12:23	12:24	12:24	12:25	12:26	12:27
2	12:30	12:31	12:32	12:33	12:34	12:35	12:36	12:38	12:39	12:40
3	12:40	12:41	12:42	12:44	12:45	12:47	12:48	12:50	12:52	12:54
4	12:49	12:51	12:53	12:54	12:56	12:58	13:00	13:02	13:05	13:07
5	12:59	13:01	13:03	13:05	13:07	13:10	13:12	13:15	13:18	13:21
6	13:09	13:11	13:13	13:16	13:19	13:22	13:25	13:28	13:31	13:35
7	13:18	13:21	13:24	13:27	13:30	13:33	13:37	13:40	13:44	13:48
8	13:28	13:31	13:35	13:38	13:42	13:45	13:49	13:53	13:58	14:02
9	13:38	13:42	13:45	13:49	13:53	13:57	14:02	14:06	14:11	14:17
10	13:48	13:52	13:56	14:00	14:05	14:10	14:15	14:20	14:25	14:31
11	13:59	14:03	14:07	14:12	14:17	14:22	14:27	14:33	14:39	14:46
12	14:09	14:14	14:18	14:24	14:29	14:35	14:41	14:47	14:54	15:01
13	14:19	14:25	14:30	14:35	14:41	14:47	14:54	15:01	15:08	15:16
14	14:30	14:36	14:41	14:47	14:54	15:00	15:08	15:15	15:23	15:32
15	14:41	14:47	14:53	15:00	15:07	15:14	15:22	15:30	15:38	15:48
16	14:52	14:58	15:05	15:12	15:20	15:27	15:36	15:45	15:54	16:04
17	15:03	15:10	15:17	15:25	15:33	15:41	15:50	16:00	16:10	16:21
18	15:15	15:22	15:30	15:38	15:47	15:56	16:06	16:16	16:27	16:39
19	15:27	15:34	15:43	15:52	16:01	16:11	16:21	16:32	16:45	16:58
20	15:39	15:47	15:56	16:05	16:15	16:26	16:37	16:49	17:03	17:17
21	15:51	16:00	16:10	16:20	16:30	16:42	16:54	17:07	17:22	17:37
22	16:04	16:13	16:24	16:34	16:46	16:58	17:12	17:26	17:41	17:58
23	16:17	16:27	16:38	16:50	17:02	17:16	17:30	17:45	18:02	18:21
24	16:31	16:42	16:53	17:06	17:19	17:34	17:49	18:06	18:25	18:45

Day length (h:m) from sunrise to sunset, lower limb on horizon.

DEC (°)	LAT = 50°	51°	52°	53°	54°	55°	56°	57°	58°	59°
-24	7:49	7:39	7:28	7:16	7:04	6:51	6:37	6:21	6:05	5:46
-23	8:02	7:53	7:43	7:32	7:20	7:08	6:55	6:41	6:25	6:09
-22	8:15	8:06	7:57	7:46	7:36	7:24	7:12	6:59	6:45	6:30
-21	8:27	8:19	8:10	8:01	7:51	7:40	7:29	7:17	7:04	6:50
-20	8:39	8:32	8:23	8:15	8:05	7:55	7:45	7:34	7:22	7:09
-19	8:51	8:44	8:36	8:28	8:19	8:10	8:00	7:50	7:39	7:27
-18	9:03	8:56	8:49	8:41	8:33	8:25	8:16	8:06	7:56	7:45
-17	9:14	9:07	9:01	8:54	8:46	8:38	8:30	8:21	8:12	8:02
-16	9:25	9:19	9:13	9:06	8:59	8:52	8:44	8:36	8:28	8:18
-15	9:36	9:30	9:24	9:18	9:12	9:05	8:58	8:51	8:43	8:34
-14	9:46	9:41	9:36	9:30	9:25	9:18	9:12	9:05	8:58	8:50
-13	9:57	9:52	9:47	9:42	9:37	9:31	9:25	9:19	9:12	9:05
-12	10:07	10:03	9:58	9:54	9:49	9:44	9:38	9:33	9:27	9:20
-11	10:17	10:13	10:09	10:05	10:01	9:56	9:51	9:46	9:41	9:35
-10	10:27	10:24	10:20	10:17	10:13	10:08	10:04	9:59	9:54	9:49
-9	10:37	10:34	10:31	10:28	10:24	10:20	10:16	10:12	10:08	10:03
-8	10:47	10:45	10:42	10:39	10:36	10:32	10:29	10:25	10:21	10:17
-7	10:57	10:55	10:52	10:50	10:47	10:44	10:41	10:38	10:35	10:31
-6	11:07	11:05	11:03	11:01	10:58	10:56	10:53	10:51	10:48	10:45
-5	11:16	11:15	11:13	11:11	11:09	11:07	11:05	11:03	11:01	10:58
-4	11:26	11:25	11:23	11:22	11:21	11:19	11:17	11:16	11:14	11:12
-3	11:36	11:35	11:34	11:33	11:32	11:30	11:29	11:28	11:27	11:25
-2	11:45	11:45	11:44	11:43	11:43	11:42	11:41	11:40	11:40	11:39
-1	11:55	11:55	11:54	11:54	11:54	11:53	11:53	11:53	11:52	11:52
0	12:04	12:04	12:04	12:05	12:05	12:05	12:05	12:05	12:05	12:05
1	12:14	12:14	12:15	12:15	12:16	12:16	12:17	12:17	12:18	12:19
2	12:23	12:24	12:25	12:26	12:27	12:28	12:29	12:30	12:31	12:32
3	12:33	12:34	12:35	12:37	12:38	12:39	12:41	12:42	12:44	12:45
4	12:43	12:44	12:46	12:47	12:49	12:51	12:53	12:55	12:57	12:59
5	12:52	12:54	12:56	12:58	13:00	13:02	13:05	13:07	13:10	13:12
6	13:02	13:04	13:06	13:09	13:11	13:14	13:17	13:20	13:23	13:26
7	13:12	13:14	13:17	13:20	13:23	13:26	13:29	13:32	13:36	13:40
8	13:22	13:24	13:28	13:31	13:34	13:38	13:41	13:45	13:49	13:54
9	13:31	13:35	13:38	13:42	13:46	13:50	13:54	13:58	14:03	14:08
10	13:42	13:45	13:49	13:53	13:57	14:02	14:06	14:11	14:17	14:22
11	13:52	13:56	14:00	14:04	14:09	14:14	14:19	14:25	14:31	14:37
12	14:02	14:06	14:11	14:16	14:21	14:27	14:32	14:38	14:45	14:52
13	14:12	14:17	14:22	14:28	14:33	14:39	14:46	14:52	14:59	15:07
14	14:23	14:28	14:34	14:40	14:46	14:52	14:59	15:06	15:14	15:22
15	14:34	14:39	14:45	14:52	14:58	15:05	15:13	15:21	15:29	15:38
16	14:45	14:51	14:57	15:04	15:11	15:19	15:27	15:36	15:45	15:54
17	14:56	15:02	15:09	15:17	15:25	15:33	15:41	15:51	16:01	16:11
18	15:07	15:14	15:22	15:30	15:38	15:47	15:56	16:06	16:17	16:29
19	15:19	15:26	15:34	15:43	15:52	16:02	16:12	16:23	16:34	16:47
20	15:31	15:39	15:48	15:57	16:06	16:17	16:28	16:39	16:52	17:05
21	15:43	15:52	16:01	16:11	16:21	16:32	16:44	16:57	17:10	17:25
22	15:56	16:05	16:15	16:25	16:36	16:48	17:01	17:15	17:30	17:46
23	16:09	16:19	16:29	16:40	16:52	17:05	17:19	17:34	17:50	18:08
24	16:22	16:33	16:44	16:56	17:09	17:23	17:38	17:54	18:12	18:31

UPPER LIMB LAT = 60° - 69°

Day length (h:m) from sunrise to sunset, upper limb on horizon.

DEC (°)	LAT = 60°	61°	62°	63°	64°	65°	66°	67°	68°	69°
-24	5:40	5:18	4:54	4:26	3:52	3:10	2:12			
-23	6:03	5:44	5:22	4:58	4:29	3:56	3:13	2:14		
-22	6:25	6:08	5:48	5:27	5:02	4:33	3:59	3:16	2:17	
-21	6:46	6:30	6:13	5:53	5:32	5:07	4:38	4:03	3:20	2:19
-20	7:06	6:52	6:36	6:18	5:59	5:37	5:12	4:43	4:08	3:24
-19	7:25	7:12	6:58	6:42	6:24	6:05	5:43	5:17	4:48	4:13
-18	7:44	7:31	7:18	7:04	6:48	6:30	6:11	5:49	5:23	4:54
-17	8:01	7:50	7:38	7:25	7:11	6:55	6:37	6:18	5:56	5:30
-16	8:18	8:08	7:57	7:45	7:32	7:18	7:02	6:45	6:25	6:03
-15	8:35	8:26	8:16	8:05	7:53	7:40	7:26	7:10	6:53	6:33
-14	8:51	8:43	8:34	8:24	8:13	8:01	7:49	7:35	7:19	7:02
-13	9:07	8:59	8:51	8:42	8:32	8:22	8:11	7:58	7:44	7:29
-12	9:23	9:16	9:08	9:00	8:51	8:42	8:32	8:21	8:08	7:55
-11	9:38	9:31	9:25	9:18	9:10	9:01	8:52	8:42	8:31	8:19
-10	9:53	9:47	9:41	9:35	9:28	9:20	9:12	9:04	8:54	8:43
-9	10:07	10:02	9:57	9:52	9:45	9:39	9:32	9:24	9:16	9:07
-8	10:22	10:18	10:13	10:08	10:03	9:57	9:51	9:44	9:37	9:29
-7	10:36	10:33	10:29	10:24	10:20	10:15	10:10	10:04	9:58	9:51
-6	10:50	10:47	10:44	10:41	10:37	10:33	10:29	10:24	10:19	10:13
-5	11:05	11:02	10:59	10:57	10:54	10:50	10:47	10:43	10:39	10:35
-4	11:19	11:17	11:15	11:13	11:10	11:08	11:05	11:02	10:59	10:56
-3	11:32	11:31	11:30	11:28	11:27	11:25	11:23	11:21	11:19	11:17
-2	11:46	11:46	11:45	11:44	11:43	11:42	11:41	11:40	11:39	11:38
-1	12:00	12:00	12:00	12:00	12:00	11:59	11:59	11:59	11:59	11:59
0	12:14	12:14	12:15	12:15	12:16	12:17	12:17	12:18	12:19	12:20
1	12:28	12:29	12:30	12:31	12:32	12:34	12:35	12:37	12:39	12:41
2	12:42	12:43	12:45	12:47	12:49	12:51	12:53	12:56	12:59	13:02
3	12:56	12:58	13:00	13:03	13:06	13:08	13:12	13:15	13:19	13:23
4	13:10	13:13	13:16	13:19	13:22	13:26	13:30	13:34	13:39	13:44
5	13:24	13:27	13:31	13:35	13:39	13:44	13:48	13:54	13:59	14:06
6	13:38	13:42	13:47	13:51	13:56	14:01	14:07	14:13	14:20	14:28
7	13:53	13:57	14:02	14:08	14:13	14:20	14:26	14:34	14:42	14:50
8	14:07	14:13	14:18	14:24	14:31	14:38	14:46	14:54	15:03	15:13
9	14:22	14:28	14:35	14:41	14:49	14:57	15:06	15:15	15:25	15:37
10	14:37	14:44	14:51	14:59	15:07	15:16	15:26	15:37	15:48	16:01
11	14:53	15:00	15:08	15:17	15:26	15:36	15:47	15:59	16:12	16:27
12	15:08	15:17	15:25	15:35	15:45	15:56	16:08	16:22	16:37	16:54
13	15:24	15:33	15:43	15:54	16:05	16:17	16:31	16:46	17:03	17:22
14	15:41	15:51	16:01	16:13	16:25	16:39	16:54	17:11	17:30	17:51
15	15:58	16:09	16:20	16:33	16:47	17:02	17:19	17:38	17:59	18:23
16	16:15	16:27	16:40	16:54	17:09	17:26	17:45	18:06	18:30	18:59
17	16:33	16:46	17:00	17:16	17:32	17:51	18:12	18:37	19:05	19:38
18	16:52	17:06	17:22	17:39	17:57	18:18	18:42	19:10	19:43	20:24
19	17:12	17:27	17:44	18:03	18:24	18:48	19:15	19:48	20:29	21:23
20	17:32	17:49	18:08	18:29	18:53	19:20	19:52	20:32	21:26	23:09
21	17:54	18:13	18:34	18:57	19:24	19:56	20:36	21:29	23:09	
22	18:17	18:38	19:01	19:28	20:00	20:39	21:32	23:10		
23	18:42	19:05	19:32	20:03	20:42	21:34	23:11			
24	19:09	19:35	20:07	20:45	21:36	23:12				

LOWER LIMB LAT = 60° - 69°

Day length (h:m) from sunrise to sunset, lower limb on horizon.

DEC (°)	LAT = 60°	61°	62°	63°	64°	65°	66°	67°	68°	69°
-24	5:26	5:03	4:37	4:06	3:29	2:41	1:23			
-23	5:50	5:30	5:07	4:40	4:10	3:32	2:43	1:24		
-22	6:13	5:55	5:34	5:11	4:44	4:13	3:36	2:46	1:25	
-21	6:34	6:18	5:59	5:39	5:15	4:49	4:17	3:39	2:49	1:27
-20	6:55	6:40	6:23	6:04	5:44	5:20	4:54	4:22	3:43	2:52
-19	7:14	7:00	6:45	6:29	6:10	5:49	5:26	4:59	4:27	3:48
-18	7:33	7:20	7:06	6:51	6:35	6:16	5:55	5:32	5:05	4:32
-17	7:51	7:39	7:27	7:13	6:58	6:41	6:23	6:02	5:38	5:11
-16	8:08	7:58	7:46	7:34	7:20	7:05	6:48	6:30	6:09	5:45
-15	8:25	8:15	8:05	7:54	7:41	7:28	7:13	6:56	6:38	6:17
-14	8:42	8:33	8:23	8:13	8:02	7:49	7:36	7:21	7:05	6:46
-13	8:58	8:49	8:41	8:31	8:21	8:10	7:58	7:45	7:30	7:14
-12	9:13	9:06	8:58	8:50	8:40	8:31	8:20	8:08	7:55	7:40
-11	9:29	9:22	9:15	9:07	8:59	8:50	8:41	8:30	8:18	8:06
-10	9:44	9:38	9:31	9:25	9:17	9:09	9:01	8:51	8:41	8:30
-9	9:58	9:53	9:48	9:42	9:35	9:28	9:21	9:12	9:03	8:54
-8	10:13	10:08	10:04	9:58	9:53	9:47	9:40	9:33	9:25	9:17
-7	10:27	10:23	10:19	10:15	10:10	10:05	9:59	9:53	9:46	9:39
-6	10:42	10:38	10:35	10:31	10:27	10:23	10:18	10:13	10:07	10:01
-5	10:56	10:53	10:50	10:47	10:44	10:40	10:36	10:32	10:28	10:23
-4	11:10	11:08	11:05	11:03	11:00	10:58	10:55	10:51	10:48	10:44
-3	11:24	11:22	11:21	11:19	11:17	11:15	11:13	11:10	11:08	11:05
-2	11:38	11:37	11:36	11:35	11:33	11:32	11:31	11:29	11:28	11:26
-1	11:52	11:51	11:51	11:50	11:50	11:49	11:49	11:48	11:48	11:47
0	12:06	12:06	12:06	12:06	12:06	12:07	12:07	12:07	12:07	12:08
1	12:19	12:20	12:21	12:22	12:23	12:24	12:25	12:26	12:27	12:29
2	12:33	12:35	12:36	12:38	12:39	12:41	12:43	12:45	12:47	12:50
3	12:47	12:49	12:51	12:53	12:56	12:58	13:01	13:04	13:07	13:11
4	13:01	13:04	13:06	13:09	13:12	13:16	13:19	13:23	13:27	13:32
5	13:15	13:18	13:22	13:25	13:29	13:33	13:38	13:42	13:48	13:53
6	13:30	13:33	13:37	13:41	13:46	13:51	13:56	14:02	14:08	14:15
7	13:44	13:48	13:53	13:58	14:03	14:09	14:15	14:22	14:29	14:37
8	13:58	14:03	14:09	14:14	14:21	14:27	14:34	14:42	14:51	15:00
9	14:13	14:19	14:25	14:31	14:38	14:46	14:54	15:03	15:13	15:24
10	14:28	14:34	14:41	14:49	14:56	15:05	15:14	15:24	15:35	15:48
11	14:43	14:50	14:58	15:06	15:15	15:24	15:35	15:46	15:59	16:13
12	14:59	15:07	15:15	15:24	15:34	15:44	15:56	16:09	16:23	16:39
13	15:15	15:23	15:33	15:43	15:53	16:05	16:18	16:32	16:48	17:06
14	15:31	15:40	15:51	16:02	16:14	16:27	16:41	16:57	17:15	17:35
15	15:48	15:58	16:09	16:21	16:34	16:49	17:05	17:23	17:43	18:05
16	16:05	16:16	16:28	16:42	16:56	17:12	17:30	17:50	18:13	18:39
17	16:23	16:35	16:48	17:03	17:19	17:37	17:57	18:19	18:45	19:16
18	16:41	16:55	17:09	17:25	17:43	18:03	18:26	18:51	19:22	19:58
19	17:00	17:15	17:31	17:49	18:09	18:31	18:57	19:27	20:03	20:49
20	17:20	17:36	17:54	18:14	18:36	19:02	19:31	20:07	20:53	22:00
21	17:41	17:59	18:19	18:41	19:06	19:36	20:11	20:56	22:02	
22	18:04	18:23	18:45	19:11	19:40	20:15	20:59	22:05		
23	18:28	18:50	19:15	19:44	20:18	21:02	22:06			
24	18:53	19:18	19:47	20:21	21:05	22:08				

UPPER LIMB LAT = 70° - 79°

Day length (h:m) from sunrise to sunset, upper limb on horizon.

DEC (°)	LAT = 70°	71°	72°	73°	74°	75°	76°	77°	78°	79°
-24										
-23										
-22										
-21										
-20	2:22									
-19	3:28	2:25								
-18	4:18	3:33	2:29							
-17	5:00	4:24	3:38	2:32						
-16	5:37	5:07	4:30	3:44	2:37					
-15	6:11	5:45	5:15	4:38	3:50	2:41				
-14	6:42	6:20	5:54	5:23	4:46	3:57	2:47			
-13	7:12	6:52	6:30	6:04	5:33	4:54	4:05	2:52		
-12	7:40	7:22	7:03	6:41	6:14	5:43	5:04	4:13	2:59	
-11	8:06	7:51	7:34	7:15	6:53	6:26	5:55	5:15	4:23	3:07
-10	8:32	8:19	8:04	7:47	7:28	7:06	6:40	6:08	5:28	4:35
-9	8:56	8:45	8:33	8:18	8:02	7:43	7:21	6:55	6:23	5:43
-8	9:21	9:11	9:00	8:48	8:34	8:18	8:00	7:39	7:13	6:41
-7	9:44	9:36	9:27	9:17	9:05	8:52	8:37	8:19	7:58	7:33
-6	10:07	10:00	9:53	9:45	9:35	9:25	9:12	8:58	8:41	8:21
-5	10:30	10:25	10:19	10:12	10:05	9:56	9:46	9:35	9:22	9:07
-4	10:52	10:48	10:44	10:39	10:33	10:27	10:20	10:11	10:02	9:50
-3	11:14	11:12	11:09	11:05	11:02	10:57	10:52	10:47	10:40	10:33
-2	11:37	11:35	11:33	11:32	11:30	11:27	11:25	11:22	11:18	11:14
-1	11:59	11:58	11:58	11:58	11:58	11:57	11:57	11:57	11:56	11:56
0	12:21	12:22	12:23	12:24	12:26	12:27	12:29	12:31	12:34	12:37
1	12:43	12:45	12:47	12:50	12:54	12:57	13:01	13:06	13:12	13:18
2	13:05	13:08	13:12	13:17	13:22	13:27	13:34	13:41	13:50	14:01
3	13:27	13:32	13:37	13:44	13:50	13:58	14:07	14:17	14:29	14:44
4	13:50	13:56	14:03	14:11	14:19	14:29	14:41	14:54	15:10	15:29
5	14:13	14:20	14:29	14:38	14:49	15:02	15:16	15:33	15:52	16:16
6	14:36	14:45	14:55	15:07	15:20	15:35	15:52	16:13	16:37	17:07
7	15:00	15:11	15:23	15:36	15:52	16:10	16:30	16:55	17:26	18:04
8	15:24	15:37	15:51	16:07	16:25	16:46	17:11	17:41	18:19	19:08
9	15:50	16:04	16:20	16:39	17:00	17:25	17:55	18:32	19:21	20:29
10	16:16	16:32	16:51	17:12	17:37	18:07	18:44	19:31	20:38	22:51
11	16:43	17:02	17:24	17:49	18:18	18:54	19:41	20:46	22:54	
12	17:12	17:34	17:59	18:28	19:04	19:49	20:52	22:56		
13	17:43	18:08	18:37	19:12	19:57	20:58	22:58			
14	18:16	18:45	19:19	20:03	21:04	23:00				
15	18:52	19:26	20:09	21:08	23:02					
16	19:32	20:15	21:13	23:04						
17	20:20	21:17	23:05							
18	21:20	23:06								
19	23:07									
20										
21										
22										
23										
24										

LOWER LIMB LAT = 70° - 79°

Day length (h:m) from sunrise to sunset, lower limb on horizon.

DEC (°)	LAT = 70°	71°	72°	73°	74°	75°	76°	77°	78°	79°
-24										
-23										
-22										
-21										
-20	1:29									
-19	2:56	1:31								
-18	3:52	3:00	1:33							
-17	4:38	3:58	3:04	1:35						
-16	5:18	4:44	4:03	3:09	1:38					
-15	5:53	5:25	4:52	4:10	3:14	1:41				
-14	6:25	6:01	5:33	4:59	4:17	3:20	1:44			
-13	6:56	6:35	6:11	5:42	5:08	4:25	3:26	1:48		
-12	7:24	7:06	6:45	6:21	5:52	5:18	4:34	3:34	1:52	
-11	7:51	7:35	7:17	6:56	6:32	6:03	5:28	4:44	3:42	1:57
-10	8:18	8:04	7:48	7:30	7:09	6:45	6:16	5:40	4:55	3:52
-9	8:43	8:31	8:17	8:01	7:44	7:23	6:59	6:30	5:54	5:08
-8	9:07	8:57	8:45	8:32	8:17	7:59	7:39	7:15	6:46	6:10
-7	9:31	9:22	9:12	9:01	8:48	8:34	8:17	7:57	7:34	7:05
-6	9:54	9:47	9:38	9:29	9:19	9:07	8:53	8:37	8:18	7:55
-5	10:17	10:11	10:04	9:57	9:48	9:39	9:28	9:15	9:00	8:42
-4	10:40	10:35	10:30	10:24	10:17	10:10	10:02	9:52	9:40	9:27
-3	11:02	10:59	10:55	10:51	10:46	10:41	10:35	10:28	10:19	10:10
-2	11:24	11:22	11:20	11:17	11:14	11:11	11:07	11:03	10:58	10:52
-1	11:46	11:45	11:44	11:43	11:42	11:41	11:39	11:38	11:36	11:33
0	12:08	12:08	12:09	12:09	12:10	12:11	12:11	12:12	12:13	12:14
1	12:30	12:32	12:34	12:36	12:38	12:41	12:44	12:47	12:51	12:56
2	12:52	12:55	12:58	13:02	13:06	13:11	13:16	13:22	13:29	13:38
3	13:14	13:19	13:23	13:29	13:34	13:41	13:49	13:58	14:08	14:20
4	13:37	13:42	13:49	13:56	14:03	14:12	14:22	14:34	14:48	15:04
5	14:00	14:07	14:14	14:23	14:33	14:44	14:57	15:12	15:29	15:50
6	14:23	14:31	14:41	14:51	15:03	15:16	15:32	15:51	16:13	16:39
7	14:46	14:56	15:07	15:20	15:34	15:50	16:09	16:32	16:59	17:33
8	15:11	15:22	15:35	15:50	16:06	16:26	16:49	17:16	17:49	18:32
9	15:35	15:49	16:04	16:21	16:41	17:03	17:31	18:04	18:46	19:43
10	16:01	16:17	16:34	16:54	17:17	17:44	18:17	18:58	19:54	21:19
11	16:28	16:45	17:05	17:28	17:56	18:28	19:09	20:03	21:26	
12	16:56	17:16	17:39	18:06	18:38	19:18	20:11	21:32		
13	17:26	17:49	18:15	18:47	19:27	20:18	21:37			
14	17:57	18:24	18:56	19:34	20:25	21:41				
15	18:32	19:03	19:41	20:31	21:45					
16	19:10	19:47	20:36	21:49						
17	19:53	20:41	21:52							
18	20:45	21:55								
19	21:58									
20										
21										
22										
23										
24										

UPPER LIMB LAT = 80° - 89°

Day length (h:m) from sunrise to sunset, upper limb on horizon.

DEC (°)	LAT = 80°	81°	82°	83°	84°	85°	86°	87°	88°	89°
-24										
-23										
-22										
-21										
-20										
-19										
-18										
-17										
-16										
-15										
-14										
-13										
-12										
-11										
-10	3:16									
-9	4:48	3:26								
-8	5:59	5:04	3:38							
-7	7:01	6:19	5:22	3:53						
-6	7:56	7:25	6:43	5:45	4:12					
-5	8:48	8:24	7:54	7:13	6:14	4:37				
-4	9:36	9:19	8:58	8:29	7:50	6:52	5:10			
-3	10:24	10:12	9:58	9:40	9:15	8:40	7:44	6:00		
-2	11:10	11:04	10:56	10:47	10:34	10:17	9:50	9:05	7:27	
-1	11:55	11:54	11:54	11:53	11:51	11:49	11:46	11:42	11:32	11:04
0	12:41	12:45	12:51	12:58	13:07	13:21	13:42	14:16	15:28	20:12
1	13:26	13:36	13:48	14:04	14:26	14:56	15:44	17:10	21:19	
2	14:13	14:29	14:48	15:14	15:49	16:41	18:08	21:49		
3	15:01	15:23	15:51	16:28	17:22	18:47	22:07			
4	15:52	16:21	17:00	17:53	19:15	22:18				
5	16:46	17:25	18:17	19:36	22:27					
6	17:46	18:37	19:53	22:34						
7	18:54	20:07	22:40							
8	20:19	22:44								
9	22:48									
10										
11										
12										
13										
14										
15										
16										
17										
18										
19										
20										
21										
22										
23										
24										

LOWER LIMB LAT = 80° - 89°

Day length (h:m) from sunrise to sunset, lower limb on horizon.

DEC (°)	LAT = 80°	81°	82°	83°	84°	85°	86°	87°	88°	89°
-24										
-23										
-22										
-21										
-20										
-19										
-18										
-17										
-16										
-15										
-14										
-13										
-12										
-11										
-10	2:02									
-9	4:03	2:08								
-8	5:23	4:16	2:16							
-7	6:29	5:41	4:31	2:25						
-6	7:27	6:51	6:02	4:50	2:37					
-5	8:20	7:53	7:17	6:28	5:14	2:52				
-4	9:10	8:50	8:24	7:49	7:01	5:45	3:12			
-3	9:58	9:44	9:26	9:03	8:31	7:44	6:28	3:42		
-2	10:45	10:36	10:25	10:11	9:52	9:26	8:45	7:32	4:33	
-1	11:31	11:27	11:23	11:17	11:10	11:00	10:45	10:19	9:27	6:33
0	12:16	12:18	12:20	12:23	12:26	12:32	12:40	12:53	13:19	14:41
1	13:01	13:08	13:17	13:28	13:43	14:05	14:37	15:33	17:38	
2	13:48	14:00	14:16	14:36	15:04	15:43	16:47	18:51		
3	14:35	14:53	15:17	15:48	16:30	17:35	19:34			
4	15:24	15:50	16:22	17:06	18:10	20:02				
5	16:17	16:50	17:34	18:37	20:23					
6	17:13	17:57	18:59	20:39						
7	18:16	19:16	20:52							
8	19:30	21:03								
9	21:12									
10										
11										
12										
13										
14										
15										
16										
17										
18										
19										
20										
21										
22										
23										
24										

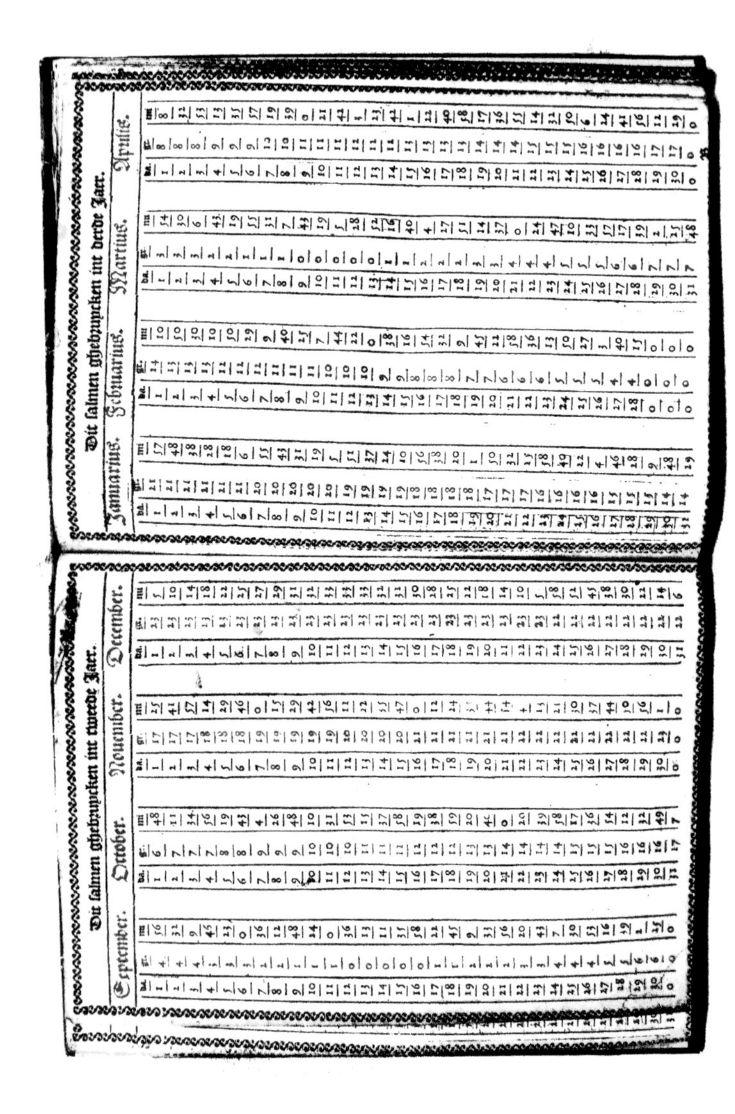

Dit salmen ghebruycken int derde Jaer.

Januarius. Februarius. Martius. Aprilis.

Dit salmen ghebruycken int tweede Jaer.

September. October. November. December.

Possibly the only surviving copy of the Graetboeck nae den Ouden Stijl (1595). In the collection of the Maritime Museum Rotterdam, inventory number 29481.

Photo: SvdW.

That the old Julian Calendar is used here, can be seen from the dates of the equinoxes, where the declination becomes zero: 14 September (second year) and 11 March (third year).

Four-year solar declination tables

The declination tables have been computed for the years 2021 -2024, using the maximum declination, the eccentricity of the Earth's orbit and the position of the apogee (where the distance Sun – Earth is maximum) and their change over time, as given by the *Astronomical Almanac* for the epoch 2000.
For each day one declination is given, computed for 12:00 GMT. They have been rounded off to integer arcminutes. Within half of an arc-minute these declinations are consistent with those of the *Nautical Almanac*, which are rounded off to one tenth of an arcminute.
Positive declinations (N) are given in red, negative declinations (S) in blue.
The idea of four-year tables is that you may use them also in later years without much loss of accuracy, because every four years the declinations of the sun nearly repeat themselves. If, for example, you would use the 2021 tables in 2025, then the maximum deviation will be 0´.7. Even twelve years later, in 2033, the error will be only 2´ at most. This maximum error occurs around the equinoxes, where the declination changes fastest.

2021

JANUARY

day	declination deg	min
1	22	58
2	22	52
3	22	46
4	22	40
5	22	33
6	22	26
7	22	18
8	22	10
9	22	2
10	21	53
11	21	43
12	21	34
13	21	23
14	21	13
15	21	2
16	20	50
17	20	39
18	20	26
19	20	14
20	20	1
21	19	47
22	19	34
23	19	20
24	19	5
25	18	50
26	18	35
27	18	20
28	18	4
29	17	48
30	17	31
31	17	14

FEBRUARY

day	declination deg	min
1	16	57
2	16	40
3	16	22
4	16	4
5	15	46
6	15	28
7	15	9
8	14	50
9	14	31
10	14	11
11	13	51
12	13	31
13	13	11
14	12	51
15	12	30
16	12	9
17	11	48
18	11	27
19	11	6
20	10	44
21	10	23
22	10	1
23	9	39
24	9	16
25	8	54
26	8	32
27	8	9
28	7	46

MARCH

day	declination deg	min
1	7	24
2	7	1
3	6	38
4	6	15
5	5	51
6	5	28
7	5	5
8	4	41
9	4	18
10	3	54
11	3	31
12	3	7
13	2	44
14	2	20
15	1	56
16	1	33
17	1	9
18	0	45
19	0	21
20	0	2
21	0	26
22	0	50
23	1	13
24	1	37
25	2	1
26	2	24
27	2	48
28	3	11
29	3	34
30	3	58
31	4	21

2021

APRIL

day	declination deg	min
1	4	44
2	5	7
3	5	30
4	5	53
5	6	16
6	6	39
7	7	1
8	7	24
9	7	46
10	8	8
11	8	30
12	8	52
13	9	14
14	9	35
15	9	57
16	10	18
17	10	39
18	11	0
19	11	21
20	11	42
21	12	2
22	12	22
23	12	42
24	13	2
25	13	21
26	13	41
27	14	0
28	14	19
29	14	37
30	14	56

MAY

day	declination deg	min
1	15	14
2	15	32
3	15	49
4	16	7
5	16	24
6	16	41
7	16	57
8	17	14
9	17	30
10	17	45
11	18	1
12	18	16
13	18	31
14	18	45
15	18	59
16	19	13
17	19	27
18	19	40
19	19	53
20	20	5
21	20	17
22	20	29
23	20	41
24	20	52
25	21	2
26	21	13
27	21	23
28	21	33
29	21	42
30	21	51
31	21	59

JUNE

day	declination deg	min
1	22	7
2	22	15
3	22	23
4	22	30
5	22	36
6	22	42
7	22	48
8	22	53
9	22	58
10	23	3
11	23	7
12	23	11
13	23	14
14	23	17
15	23	20
16	23	22
17	23	24
18	23	25
19	23	26
20	23	26
21	23	26
22	23	26
23	23	25
24	23	24
25	23	23
26	23	21
27	23	18
28	23	15
29	23	12
30	23	8

2021

JULY

day	declination deg	min
1	23	4
2	23	0
3	22	55
4	22	50
5	22	44
6	22	38
7	22	32
8	22	25
9	22	18
10	22	10
11	22	2
12	21	54
13	21	45
14	21	36
15	21	27
16	21	17
17	21	7
18	20	56
19	20	45
20	20	34
21	20	22
22	20	10
23	19	58
24	19	45
25	19	33
26	19	19
27	19	6
28	18	52
29	18	38
30	18	23
31	18	8

AUGUST

day	declination deg	min
1	17	53
2	17	38
3	17	22
4	17	6
5	16	50
6	16	33
7	16	17
8	16	0
9	15	42
10	15	25
11	15	7
12	14	49
13	14	31
14	14	12
15	13	53
16	13	35
17	13	15
18	12	56
19	12	36
20	12	17
21	11	57
22	11	36
23	11	16
24	10	56
25	10	35
26	10	14
27	9	53
28	9	32
29	9	11
30	8	49
31	8	27

SEPTEMBER

day	declination deg	min
1	8	6
2	7	44
3	7	22
4	7	0
5	6	37
6	6	15
7	5	53
8	5	30
9	5	7
10	4	45
11	4	22
12	3	59
13	3	36
14	3	13
15	2	50
16	2	27
17	2	4
18	1	40
19	1	17
20	0	54
21	0	30
22	0	7
23	0	16
24	0	40
25	1	3
26	1	26
27	1	50
28	2	13
29	2	36
30	3	0

2021

OCTOBER

day	declination deg	min
1	3	23
2	3	46
3	4	9
4	4	32
5	4	56
6	5	19
7	5	42
8	6	4
9	6	27
10	6	50
11	7	13
12	7	35
13	7	58
14	8	20
15	8	42
16	9	4
17	9	26
18	9	48
19	10	10
20	10	31
21	10	52
22	11	14
23	11	35
24	11	56
25	12	16
26	12	37
27	12	57
28	13	17
29	13	37
30	13	57
31	14	16

NOVEMBER

day	declination deg	min
1	14	35
2	14	54
3	15	13
4	15	32
5	15	50
6	16	8
7	16	25
8	16	43
9	17	0
10	17	17
11	17	33
12	17	50
13	18	6
14	18	21
15	18	37
16	18	52
17	19	6
18	19	21
19	19	34
20	19	48
21	20	1
22	20	14
23	20	27
24	20	39
25	20	51
26	21	2
27	21	13
28	21	23
29	21	34
30	21	43

DECEMBER

day	declination deg	min
1	21	53
2	22	2
3	22	10
4	22	18
5	22	26
6	22	33
7	22	40
8	22	46
9	22	52
10	22	57
11	23	2
12	23	7
13	23	11
14	23	14
15	23	18
16	23	20
17	23	22
18	23	24
19	23	25
20	23	26
21	23	26
22	23	26
23	23	26
24	23	25
25	23	23
26	23	21
27	23	18
28	23	15
29	23	12
30	23	8
31	23	4

2022

JANUARY

day	declination deg	min
1	22	59
2	22	54
3	22	48
4	22	42
5	22	35
6	22	28
7	22	20
8	22	12
9	22	4
10	21	55
11	21	46
12	21	36
13	21	26
14	21	15
15	21	5
16	20	53
17	20	41
18	20	29
19	20	17
20	20	4
21	19	51
22	19	37
23	19	23
24	19	9
25	18	54
26	18	39
27	18	23
28	18	8
29	17	52
30	17	35
31	17	19

FEBRUARY

day	declination deg	min
1	17	2
2	16	44
3	16	27
4	16	9
5	15	51
6	15	32
7	15	13
8	14	55
9	14	35
10	14	16
11	13	56
12	13	36
13	13	16
14	12	56
15	12	35
16	12	14
17	11	54
18	11	32
19	11	11
20	10	50
21	10	28
22	10	6
23	9	44
24	9	22
25	9	0
26	8	37
27	8	15
28	7	52

MARCH

day	declination deg	min
1	7	29
2	7	6
3	6	43
4	6	20
5	5	57
6	5	34
7	5	11
8	4	47
9	4	24
10	4	0
11	3	37
12	3	13
13	2	49
14	2	26
15	2	2
16	1	38
17	1	15
18	0	51
19	0	27
20	0	3
21	0	20
22	0	44
23	1	8
24	1	31
25	1	55
26	2	18
27	2	42
28	3	5
29	3	29
30	3	52
31	4	15

2022

APRIL

day	declination deg	min
1	4	39
2	5	2
3	5	25
4	5	48
5	6	10
6	6	33
7	6	56
8	7	18
9	7	41
10	8	3
11	8	25
12	8	47
13	9	9
14	9	30
15	9	52
16	10	13
17	10	34
18	10	55
19	11	16
20	11	37
21	11	57
22	12	17
23	12	37
24	12	57
25	13	17
26	13	36
27	13	55
28	14	14
29	14	33
30	14	51

MAY

day	declination deg	min
1	15	10
2	15	27
3	15	45
4	16	3
5	16	20
6	16	37
7	16	53
8	17	10
9	17	26
10	17	42
11	17	57
12	18	12
13	18	27
14	18	42
15	18	56
16	19	10
17	19	23
18	19	37
19	19	50
20	20	2
21	20	14
22	20	26
23	20	38
24	20	49
25	21	0
26	21	10
27	21	21
28	21	30
29	21	40
30	21	49
31	21	57

JUNE

day	declination deg	min
1	22	5
2	22	13
3	22	21
4	22	28
5	22	35
6	22	41
7	22	47
8	22	52
9	22	57
10	23	2
11	23	6
12	23	10
13	23	14
14	23	17
15	23	19
16	23	21
17	23	23
18	23	25
19	23	26
20	23	26
21	23	26
22	23	26
23	23	26
24	23	24
25	23	23
26	23	21
27	23	19
28	23	16
29	23	13
30	23	9

2022

JULY

day	declination deg	min
1	23	5
2	23	1
3	22	56
4	22	51
5	22	46
6	22	40
7	22	33
8	22	27
9	22	20
10	22	12
11	22	4
12	21	56
13	21	47
14	21	38
15	21	29
16	21	19
17	21	9
18	20	59
19	20	48
20	20	37
21	20	25
22	20	13
23	20	1
24	19	49
25	19	36
26	19	23
27	19	9
28	18	55
29	18	41
30	18	27
31	18	12

AUGUST

day	declination deg	min
1	17	57
2	17	42
3	17	26
4	17	10
5	16	54
6	16	37
7	16	21
8	16	4
9	15	47
10	15	29
11	15	11
12	14	53
13	14	35
14	14	17
15	13	58
16	13	39
17	13	20
18	13	1
19	12	41
20	12	21
21	12	2
22	11	41
23	11	21
24	11	1
25	10	40
26	10	19
27	9	58
28	9	37
29	9	16
30	8	54
31	8	33

SEPTEMBER

day	declination deg	min
1	8	11
2	7	49
3	7	27
4	7	5
5	6	43
6	6	20
7	5	58
8	5	36
9	5	13
10	4	50
11	4	27
12	4	5
13	3	42
14	3	19
15	2	56
16	2	32
17	2	9
18	1	46
19	1	23
20	0	59
21	0	36
22	0	13
23	0	11
24	0	34
25	0	57
26	1	21
27	1	44
28	2	7
29	2	31
30	2	54

2022

OCTOBER

day	declination deg	min
1	3	17
2	3	41
3	4	4
4	4	27
5	4	50
6	5	13
7	5	36
8	5	59
9	6	22
10	6	44
11	7	7
12	7	30
13	7	52
14	8	14
15	8	37
16	8	59
17	9	21
18	9	43
19	10	4
20	10	26
21	10	47
22	11	9
23	11	30
24	11	51
25	12	11
26	12	32
27	12	52
28	13	12
29	13	32
30	13	52
31	14	11

NOVEMBER

day	declination deg	min
1	14	31
2	14	50
3	15	9
4	15	27
5	15	45
6	16	3
7	16	21
8	16	39
9	16	56
10	17	13
11	17	29
12	17	46
13	18	2
14	18	18
15	18	33
16	18	48
17	19	3
18	19	17
19	19	31
20	19	45
21	19	58
22	20	11
23	20	24
24	20	36
25	20	48
26	20	59
27	21	10
28	21	21
29	21	31
30	21	41

DECEMBER

day	declination deg	min
1	21	50
2	22	0
3	22	8
4	22	16
5	22	24
6	22	31
7	22	38
8	22	45
9	22	51
10	22	56
11	23	1
12	23	6
13	23	10
14	23	14
15	23	17
16	23	20
17	23	22
18	23	24
19	23	25
20	23	26
21	23	26
22	23	26
23	23	26
24	23	25
25	23	23
26	23	21
27	23	19
28	23	16
29	23	13
30	23	9
31	23	5

2023

JANUARY

day	declination deg	min
1	23	0
2	22	55
3	22	49
4	22	43
5	22	37
6	22	30
7	22	22
8	22	14
9	22	6
10	21	57
11	21	48
12	21	38
13	21	28
14	21	18
15	21	7
16	20	56
17	20	44
18	20	32
19	20	20
20	20	7
21	19	54
22	19	40
23	19	26
24	19	12
25	18	57
26	18	42
27	18	27
28	18	12
29	17	56
30	17	39
31	17	23

FEBRUARY

day	declination deg	min
1	17	6
2	16	48
3	16	31
4	16	13
5	15	55
6	15	37
7	15	18
8	14	59
9	14	40
10	14	21
11	14	1
12	13	41
13	13	21
14	13	1
15	12	40
16	12	20
17	11	59
18	11	38
19	11	16
20	10	55
21	10	33
22	10	11
23	9	49
24	9	27
25	9	5
26	8	43
27	8	20
28	7	57

MARCH

day	declination deg	min
1	7	35
2	7	12
3	6	49
4	6	26
5	6	3
6	5	39
7	5	16
8	4	53
9	4	29
10	4	6
11	3	42
12	3	19
13	2	55
14	2	31
15	2	8
16	1	44
17	1	20
18	0	57
19	0	33
20	0	9
21	0	15
22	0	38
23	1	2
24	1	26
25	1	49
26	2	13
27	2	36
28	3	0
29	3	23
30	3	46
31	4	10

2023

APRIL

day	declination deg	min
1	4	33
2	4	56
3	5	19
4	5	42
5	6	5
6	6	28
7	6	50
8	7	13
9	7	35
10	7	57
11	8	19
12	8	41
13	9	3
14	9	25
15	9	47
16	10	8
17	10	29
18	10	50
19	11	11
20	11	32
21	11	52
22	12	12
23	12	32
24	12	52
25	13	12
26	13	31
27	13	51
28	14	10
29	14	28
30	14	47

MAY

day	declination deg	min
1	15	5
2	15	23
3	15	41
4	15	58
5	16	16
6	16	33
7	16	49
8	17	6
9	17	22
10	17	38
11	17	53
12	18	9
13	18	23
14	18	38
15	18	52
16	19	6
17	19	20
18	19	33
19	19	46
20	19	59
21	20	11
22	20	23
23	20	35
24	20	46
25	20	57
26	21	8
27	21	18
28	21	28
29	21	37
30	21	46
31	21	55

JUNE

day	declination deg	min
1	22	4
2	22	11
3	22	19
4	22	26
5	22	33
6	22	39
7	22	45
8	22	51
9	22	56
10	23	1
11	23	5
12	23	9
13	23	13
14	23	16
15	23	19
16	23	21
17	23	23
18	23	24
19	23	25
20	23	26
21	23	26
22	23	26
23	23	26
24	23	25
25	23	23
26	23	22
27	23	19
28	23	17
29	23	14
30	23	10

2023

JULY

day	declination deg	min
1	23	6
2	23	2
3	22	58
4	22	52
5	22	47
6	22	41
7	22	35
8	22	28
9	22	21
10	22	14
11	22	6
12	21	58
13	21	49
14	21	40
15	21	31
16	21	22
17	21	12
18	21	1
19	20	50
20	20	39
21	20	28
22	20	16
23	20	4
24	19	52
25	19	39
26	19	26
27	19	12
28	18	59
29	18	45
30	18	30
31	18	16

AUGUST

day	declination deg	min
1	18	1
2	17	45
3	17	30
4	17	14
5	16	58
6	16	41
7	16	25
8	16	8
9	15	51
10	15	33
11	15	16
12	14	58
13	14	40
14	14	21
15	14	3
16	13	44
17	13	25
18	13	5
19	12	46
20	12	26
21	12	6
22	11	46
23	11	26
24	11	6
25	10	45
26	10	24
27	10	3
28	9	42
29	9	21
30	8	59
31	8	38

SEPTEMBER

day	declination deg	min
1	8	16
2	7	54
3	7	32
4	7	10
5	6	48
6	6	26
7	6	3
8	5	41
9	5	18
10	4	56
11	4	33
12	4	10
13	3	47
14	3	24
15	3	1
16	2	38
17	2	15
18	1	52
19	1	28
20	1	5
21	0	42
22	0	18
23	0	5
24	0	28
25	0	52
26	1	15
27	1	38
28	2	2
29	2	25
30	2	48

2023

OCTOBER

day	declination deg	min
1	3	12
2	3	35
3	3	58
4	4	21
5	4	44
6	5	7
7	5	30
8	5	53
9	6	16
10	6	39
11	7	2
12	7	24
13	7	47
14	8	9
15	8	31
16	8	53
17	9	15
18	9	37
19	9	59
20	10	21
21	10	42
22	11	3
23	11	25
24	11	45
25	12	6
26	12	27
27	12	47
28	13	7
29	13	27
30	13	47
31	14	7

NOVEMBER

day	declination deg	min
1	14	26
2	14	45
3	15	4
4	15	23
5	15	41
6	15	59
7	16	17
8	16	34
9	16	52
10	17	9
11	17	25
12	17	42
13	17	58
14	18	14
15	18	29
16	18	44
17	18	59
18	19	14
19	19	28
20	19	42
21	19	55
22	20	8
23	20	21
24	20	33
25	20	45
26	20	56
27	21	8
28	21	18
29	21	29
30	21	39

DECEMBER

day	declination deg	min
1	21	48
2	21	57
3	22	6
4	22	14
5	22	22
6	22	30
7	22	37
8	22	43
9	22	49
10	22	55
11	23	0
12	23	5
13	23	9
14	23	13
15	23	16
16	23	19
17	23	21
18	23	23
19	23	25
20	23	26
21	23	26
22	23	26
23	23	26
24	23	25
25	23	24
26	23	22
27	23	20
28	23	17
29	23	14
30	23	10
31	23	6

2024

JANUARY

day	declination deg	min
1	23	1
2	22	56
3	22	51
4	22	45
5	22	38
6	22	31
7	22	24
8	22	16
9	22	8
10	21	59
11	21	50
12	21	41
13	21	31
14	21	21
15	21	10
16	20	59
17	20	47
18	20	35
19	20	23
20	20	10
21	19	57
22	19	44
23	19	30
24	19	16
25	19	1
26	18	46
27	18	31
28	18	15
29	17	59
30	17	43
31	17	27

FEBRUARY

day	declination deg	min
1	17	10
2	16	53
3	16	35
4	16	17
5	15	59
6	15	41
7	15	23
8	15	4
9	14	45
10	14	25
11	14	6
12	13	46
13	13	26
14	13	6
15	12	45
16	12	25
17	12	4
18	11	43
19	11	21
20	11	0
21	10	38
22	10	17
23	9	55
24	9	33
25	9	10
26	8	48
27	8	26
28	8	3
29	7	40

MARCH

day	declination deg	min
1	7	17
2	6	54
3	6	31
4	6	8
5	5	45
6	5	22
7	4	58
8	4	35
9	4	12
10	3	48
11	3	24
12	3	1
13	2	37
14	2	14
15	1	50
16	1	26
17	1	2
18	0	39
19	0	15
20	0	9
21	0	33
22	0	56
23	1	20
24	1	43
25	2	7
26	2	31
27	2	54
28	3	17
29	3	41
30	4	4
31	4	27

2024

APRIL

day	declination deg	min
1	4	50
2	5	14
3	5	36
4	5	59
5	6	22
6	6	45
7	7	7
8	7	30
9	7	52
10	8	14
11	8	36
12	8	58
13	9	20
14	9	41
15	10	3
16	10	24
17	10	45
18	11	6
19	11	27
20	11	47
21	12	7
22	12	28
23	12	47
24	13	7
25	13	27
26	13	46
27	14	5
28	14	24
29	14	42
30	15	1

MAY

day	declination deg	min
1	15	19
2	15	37
3	15	54
4	16	12
5	16	29
6	16	45
7	17	2
8	17	18
9	17	34
10	17	50
11	18	5
12	18	20
13	18	35
14	18	49
15	19	3
16	19	17
17	19	30
18	19	43
19	19	56
20	20	9
21	20	21
22	20	32
23	20	44
24	20	55
25	21	5
26	21	16
27	21	26
28	21	35
29	21	44
30	21	53
31	22	2

JUNE

day	declination deg	min
1	22	10
2	22	17
3	22	25
4	22	31
5	22	38
6	22	44
7	22	50
8	22	55
9	23	0
10	23	4
11	23	8
12	23	12
13	23	15
14	23	18
15	23	20
16	23	22
17	23	24
18	23	25
19	23	26
20	23	26
21	23	26
22	23	26
23	23	25
24	23	24
25	23	22
26	23	20
27	23	17
28	23	14
29	23	11
30	23	7

2024

JULY

day	declination deg	min
1	23	3
2	22	59
3	22	54
4	22	48
5	22	43
6	22	36
7	22	30
8	22	23
9	22	16
10	22	8
11	22	0
12	21	52
13	21	43
14	21	33
15	21	24
16	21	14
17	21	4
18	20	53
19	20	42
20	20	31
21	20	19
22	20	7
23	19	55
24	19	42
25	19	29
26	19	16
27	19	2
28	18	48
29	18	34
30	18	19
31	18	4

AUGUST

day	declination deg	min
1	17	49
2	17	34
3	17	18
4	17	2
5	16	45
6	16	29
7	16	12
8	15	55
9	15	38
10	15	20
11	15	2
12	14	44
13	14	26
14	14	7
15	13	48
16	13	29
17	13	10
18	12	51
19	12	31
20	12	11
21	11	51
22	11	31
23	11	11
24	10	50
25	10	29
26	10	8
27	9	47
28	9	26
29	9	5
30	8	43
31	8	21

SEPTEMBER

day	declination deg	min
1	8	0
2	7	38
3	7	16
4	6	54
5	6	31
6	6	9
7	5	46
8	5	24
9	5	1
10	4	38
11	4	16
12	3	53
13	3	30
14	3	7
15	2	44
16	2	20
17	1	57
18	1	34
19	1	11
20	0	47
21	0	24
22	0	1
23	0	23
24	0	46
25	1	9
26	1	33
27	1	56
28	2	19
29	2	43
30	3	6

2024

OCTOBER

day	declination deg	min
1	3	29
2	3	53
3	4	16
4	4	39
5	5	2
6	5	25
7	5	48
8	6	11
9	6	33
10	6	56
11	7	19
12	7	41
13	8	4
14	8	26
15	8	48
16	9	10
17	9	32
18	9	54
19	10	15
20	10	37
21	10	58
22	11	19
23	11	40
24	12	1
25	12	22
26	12	42
27	13	3
28	13	23
29	13	42
30	14	2
31	14	21

NOVEMBER

day	declination deg	min
1	14	41
2	14	59
3	15	18
4	15	37
5	15	55
6	16	13
7	16	30
8	16	48
9	17	5
10	17	21
11	17	38
12	17	54
13	18	10
14	18	25
15	18	41
16	18	56
17	19	10
18	19	24
19	19	38
20	19	52
21	20	5
22	20	18
23	20	30
24	20	42
25	20	54
26	21	5
27	21	16
28	21	26
29	21	36
30	21	46

DECEMBER

day	declination deg	min
1	21	55
2	22	4
3	22	12
4	22	20
5	22	28
6	22	35
7	22	42
8	22	48
9	22	53
10	22	59
11	23	4
12	23	8
13	23	12
14	23	15
15	23	18
16	23	21
17	23	23
18	23	24
19	23	26
20	23	26
21	23	26
22	23	26
23	23	25
24	23	24
25	23	22
26	23	20
27	23	18
28	23	15
29	23	11
30	23	7
31	23	2

Bibliography and sources

Astronomical

Toomer, G.J., Ptolemy's Almagest, Princeton University Press, Princeton, New Jersey, 1998.
Danck de Saxonia, J., Tabulae astronomicae Alphonsinae, Venice, 1483.
Santritter, J.L., Tabule astronomice Alfonsi Regis, Venice, 1492.
Zacut, A., Almanach Perpetuus coelestius motuus, translated from Hebrew by José Vizinho, Leiria, 1496.
Regiomontanus, J., Tabulae directionum profectionumque, Ed. Johannes Engel, Augsburg, 1490.
Regiomontanus, J., Ephemerides 1474 – 1506, Nuremberg, 1474. Reprint: Ephemerides 1492 – 1506, Augsburg, 1492.
Copernicus, N., De revolutionibus orbium coelestium libri VI, Nuremberg, 1543.
Reinhold, E. 1551, Prutenicae Tabulae Coelestium Motuum, Tübingen, 1551.
Stadius, J., Ephemerides novae, for 1554 – 1606 many editions, printed in Cologne and Lyon.
Brahe, T., Astronomiae Instauratae Progymnasmata, Prague, 1602.

Nautical, in part later facsimiles

Bensaúde, J., L' Astronomie nautique au Portugal a l'époque des grandes découvertes, Bern, 1912.
Bensaúde, J., Regimento do estrolabio e do quadrante : tractado da spera do mundo : reproduction fac-similé du seul exemplaire connu appartenant à la Bibliothèque Royale de Munich, Munich, 1914.
Albuquerque, L. M. de, Os guias náuticos de Munique e Évora, Junta de Investigacões do Ultramar, Lisbon, 1965.
Lisboa, J. de, Livro de Marinharia, Lisbon, c. 1516.
Fernandes, V., Reportório dos tempos, Lisbon, 1518.
Pires, A., Livro de Marinharia, Bibliothèque Nat. de Paris, Codice 44340 (Ms. Port., 40), pp. 1-37.
Albuquerque, L. M. de, O livro de marinharia de André Pires, , Junta de Investigacões do Ultramar, Lisbon, 1963.
Enciso, M.F. de, Suma de Geographia, Seville, 1519.
Enciso, M.F. de, Suma de Geographia, Seville, 1530.
Faleiro, F. 1535, Tratado del esphera y del arte del marear, Seville, 1535.
Medina, P. de, Arte de Nauegar, Valladolid, 1545.

Medina, P. de, L'art de Naviguer, translated and extended by N. de Nicolai, Lyon, 1554.
Medina, P. de, The Arte of Navigation, translated by John Frampton, London, 1581.
Medina, P. de, De Zeevaert oft Conste van ter Zee te varen, translated by M. Everaert and extended by M. Coignet, Antwerp, 1580.
Nunez, P., Tratrado da sphera, Lisbon, 1537.
Cortes, M., Breve Compendio de la sphera y de arte de nauegar..., Seville, 1551.
Cortes, M., The Arte of Nauigation Conteynyng a Compendious Description of the Sphere, with the Makyng of Certen Instrumentes and Rules for Nauigations ... Translated out of Spanyshe into Englyshe by Richard Eden, London, 1561.
Bourne, W., A Regiment for the Sea, London, 1573. Reprint 1577.
Bourne, W., A Regiment for the Sea, corrected and extended by T. Hood, London, 1593.
Claesz, C., Graetboecxken nae den Nieuwen Stijl, Amsterdam, 1587.
Claesz, C., Graetboeck nae den Ouden Stijl, Amsterdam, 1595.
Waghenaer, L.J., Spieghel der Zeevaerdt, Leyden, 1584.
Waghenaer, L.J., The Mariners Mirrour, London, 1588.
Waghenaer, L.J, Thresoor der Zeevaert, Leyden, 1592.
Barentsz, W., Nieuwe Beschryvinghe ende Caertboeck vande Midtlandtsche Zee, Amsterdam, 1595.
Blaeu, W. J., Het Licht der Zeevaert, Amsterdam, 1608.
Blaeu, W. J., The Light of Navigation, Amsterdam, 1612.

Miscillaneous

Veer, G. de, Waerachtighe Beschryvinghe van drie seylagiën ter werelt noyt soo vreemt ghehoort, Cornelis Claesz, Amsterdam, 1598; Latin translation: Diarium nauticum seu descriptio trium navigationum admirandarum, Cornelis Claesz, Amsterdam 1598; German translation: Warhafftige Relation. Der dreyen newen unerhörten, seltzamen Schiffart, L. Hulsius, Nuremberg 1598; French translation: Trois navigations admirables faictes par le Hollandais et Zelandois au Septentrion, G. Chaudière, Paris 1599; Italian translation: Tre navigationi fatte dagli Olandesi, e Zelandesi al settentrione, G.B. Ciotto, Venice 1599; English translation: The true and perfect description of three voyages, so strange and woonderfull that the like hath neuer been heard of before, translated by William Phillip, ed. T. Pauier, Londen 1609
Moitessier, B., La longue route, Arthaud, Paris, 1971. The Long Way, trans. William Rodarmor, Adlard Coles Nautical, UK, 1974.

The authors

Siebren van der Werf *(1942) is a retired physicist of the University of Groningen, The Netherlands. Besides sailing, his current interest and work is on the history of navigation and on refraction of light in the atmosphere and computer simulations of its anomalies, such as the fata morgana and the Novaya Zemlya phenomenon. In 1997 he received the Samuel Burka Award of the American Institute of Navigation for his study on the lunar distance method.*

Books:

Het Nova Zembla verschijnsel, Geschiedenis van een luchtspiegeling, 2011, Historische Uitgeverij, Groningen. ISBN 978-90-6554-080-5.

Astronavigatie van Columbus tot Willem Barentsz, 2017, Lanasta, Emmen. Coauthor: Dick Huges. ISBN 978-90-8616-159-1.

Journals:

Reworking Slocum's Navigation, Yachting Monthly, April (1995) 42-44.

The lunar distance method in the nineteenth century: A simulation of Joshua Slocum's observation on June 16, 1896, Navigation, Journal of the Institute of Navigation 44(1997)1-13.

Astronomical observations during Willem Barents's third voyage to the North (1596-1597), ARCTIC 51(1998)142-154.

Raytracing and refraction in the modified US1976 atmosphere, Applied Optics 42(2003)354-366

The Novaya Zemlya effect and sunsets, Applied Optics 42(2003)367-378. Coauthors G. P. Können and W. H. Lehn.

Gerrit de Veer's true and perfect description of the Novaya Zemlya effect, Applied Optics 42(2003)379-389. Coauthors G. P. Können, W. H. Lehn, F. Steenhuisen and W. P. S. Davidson.

Atmospheric refraction: a history, Applied Optics 44(2005)5624-5636. First author W.H. Lehn.

Wave Height and Horizon Dip, Navigation, Journal of The Institute of Navigation Vol. 62, No. 2, Summer 2015, 161 – 169. Coauthor Igor Shokaryev.

Hafgerdingar and giant waves, Applied Optics 56 (2017), S. G51-G58.

Hafgerdingar and giant waves, Applied Optics 56(2017)G51-G58.

History and critical analysis of fifteenth and sixteenth century nautical tables, Journal for the History of Astronomy 48(2017)207-232.

Vroege instrumenten voor breedte- en tijdbepaling, Scheepshistorie 24(2018)56-67.

Nautical Tables for Vasco da Gama, 1497-1501?, Journal for the History of Astronomy 50(2019)326-338.

Dick Huges *(1944) is a chest physician by profession (retired).*
He made several long-distances sailing journeys including a solo-circumnavigation of the world. In 2000 and 2003 he received the "Voorzittersprijs der Nederlandsche Vereeniging van Kustzeilers (the Chairman's award of the Dutch Association of Coastal Yachtsmen), and in 2005 he won the "Trans-Ocean-Medaille für hervorragende hochseeseglerische Leistungen" (T-O- Medaille for extraordinary ocean-sailing achievements). Dick's conclusion after 70.000 miles on the high seas is: "stay in touch with the basics and focus on the essentials, not on luxuries".

Books:
Zeilen met losse handen; Over koershouden met stuurzeilen. 2012, Watersportmedia. (E-book).

Klassiek Zeemanschap in de Praktijk; solo onder zeil de wereld rond, 2015, Watersportmedia. ISBN 978-94-91201-02-8.

Boordboek Zelfredzaam Zeemanschap; veilig thuiskomen na uitval elektronica, 2015, ALK/Heijnen (vanaf 2018: Hollandia). ISBN 978-90596-112-45.

De Zon als GPS; een praktische handleiding voor astronavigatie, 2016, Watersportmedia. ISBN 978-94912-010-42.

Boordboek Solo Zeemanschap; voor veilig onderbemand zeilen, 2017, ALK/Heijnen (vanaf 2018: Hollandia). ISBN 978-90-5961-134-4.

Boordboek Buitengaats Zeemanschap; veilig zeezeilen op de Noordzee, 2018, Hollandia. ISBN 978-90-6410-661-3.

Journals:
Zeezeilen: droom en werkelijkheid, Arts in Beweging 23(1994)7-10.

Lange Kielen, Lange Reizen, over S-spant langkielers, ZEILEN (Vol. 6-2010)86-91.

Zeilen met losse handen; over koers houden met stuurzeilen, ZEILEN (Vol. 4-2012)36-40.

Zon en Sterren als GPS; terug naar de basis, ZEILEN (Vol. 10-2013)80-83 .

Jacques Yves Le Toumelin: uitvinder van het Sheet to Tiller zelfstuursysteem, Spiegel der Zeilvaart (Vol. 5-2013)57-59.

Rekenen met de maan; over de bepaling van de lengtegraad zonder chronometer, Spiegel der Zeilvaart (Vol. 1-2016)30-33.

Over navigatie vroeger en nu, Scheepshistorie 21(2016)86-103, Lanasta Emmen. ISBN 978-90-8616-218-5.

Navigatie instrumenten door de eeuwen heen, Scheepshistorie 22(2016)20-31, Lanasta Emmen. ISBN 978-90-8616-219-2.

Zeekaarten, Scheepshistorie 23 (2017)62-73, Lanasta Emmen. ISBN 978-90-8616-220-8.

Meer inzicht in navigatie op zee, Spiegel der Zeilvaart (Vol.10-2019)60-63.

Van Gis naar Fix, Scheepshistorie 28 (2020)106-116, Lanasta Odoorn, ISBN: 978-90-8616-335-9

Websites of the authors:
https://siebrenvanderwerf.nl/
www.sextantnavigatie.nl
www.zelfredzaamzeemanschap.nl

THANKS!
The authors wish to thank Ann Scholten-Sampson and Hans F. Buré for reading the manuscript and for useful suggestions.

Notes

Notes

Notes

Notes